Irrigation and Drainage

Irrigation and Drainage

Guru Sharan Das

RANDOM

RANDOM PUBLICATIONS

NEW DELHI (INDIA)

Irrigation and Drainage

ISBN 978-93-5111-359-1

Published in 2014 in India by

Reprint 2020

RANDOM PUBLICATIONS
4376-A/4B, Gali Murari Lal, Ansari Road
New Delhi-110 002
Phone : +91-11-43580356, +91-11-23289044
e-mail: randomexports@gmail.com, sales@randompublications.com, info@randompublications.com

Type Setting by : Keystoneprintads, Delhi-110051
Printed at : Mehra Printers, Delhi-110 092

Preface

Irrigation and drainage, artificial application of water to land and artificial removal of excess water from land, respectively. Some land requires irrigation or drainage before it is possible to use it for any agricultural production; other land profits from either practice to increase production. Some land, of course, does not need either. Although either practice may be, and both often are, used for nonagricultural purposes to improve the environment, this article is limited to their application to agriculture.

Water is an important constituent of the plant body. Plants need water to carry on different physiological processes essential to their growth and development. A great deal of water from plant body is wasted through the process of transpiration. Therefore, to compensate this loss of water, absorption of water from soil is an important phenomenon. Soil gets water mainly from two sources viz. precipitation and irrigation. Hence, irrigation is a process, other than natural precipitation, which supplies water to crops, orchards, grass, or any other cultivated plants.

In the wetter parts of the world where rainfed cultivation is practiced, the farmer's activities consist of selecting suitable land, preparing the soil for cultivation; sowing, tending and harvesting the crops. Natural rainfall provides the water needed. But in many other places otherwise favourable for cultivation, natural rainfall does not provide all the water needed and, as such, irrigation of crops is required to make up this deficiency.

Drainage can be either natural or artificial. Many areas have some natural drainage; this means that excess water flows from the farmers' fields to swamps or to lakes and rivers. Natural drainage, however, is often inadequate and artificial or man-made drainage is required.

The book will be useful as a reference source to the students, teachers, researchers, policy makers, administrators and managers of irrigation and drainage as well as rural development.

I thank all members of my team who have helped in the preparation of the book. My special thanks go to "Random Publications" who have published the book.

– Guru Sharan Das

Contents

1

Conveyance of Irrigation System

IRRIGATION SYSTEM

The irrigation system consists of a (main) intake structure or (main) pumping station, a conveyance system, a distribution system, a field application system and a drainage system.

The (main) intake structure, or (main) pumping station, directs water from the source of supply, such as a reservoir or a river, into the irrigation system. The conveyance system assures the transport of water from the main intake structure or main pumping station up to the field ditches. The distribution system assures the transport of water through field ditches to the irrigated fields. The field application system assures the transport of water within the fields. The drainage system removes the excess water (caused by rainfall and/ or irrigation) from the fields.

FIELD APPLICATION SYSTEMS

There are many methods of applying water to the field. The simplest one consists of bringing water from the source of supply, such as a well, to each plant with a bucket or a water-can. This is a very time-consuming method and it involves quite heavy work. However, it can be used successfully to irrigate small plots of land, such as vegetable gardens, that are in the neighbourhood of a water source. More sophisticated methods of water application are used in larger irrigation systems. There are three basic methods: surface irrigation, sprinkler irrigation and drip irrigation.

SURFACE IRRIGATION

Surface irrigation is the application of water to the fields at ground level. Either the entire field is flooded or the water is directed into furrows or borders.

Furrow Irrigation

Furrows are narrow ditches dug on the field between the rows of crops. The water runs along them as it moves down the slope of the field. The water flows from the field ditch into the furrows by opening up the bank or dyke of the ditch or by means of syphons or spiles. Siphons are small curved pipes that deliver water over the ditch bank. Spiles are small pipes buried in the ditch bank.

Border Irrigation

In border irrigation, the field to be irrigated is divided into strips (also called borders or borderstrips) by parallel dykes or border ridges. The water is released from the field ditch onto the border through gate structures called outlets. The water can also be released by means of siphons or spiles. The sheet of flowing water moves down the slope of the border, guided by the border ridges.

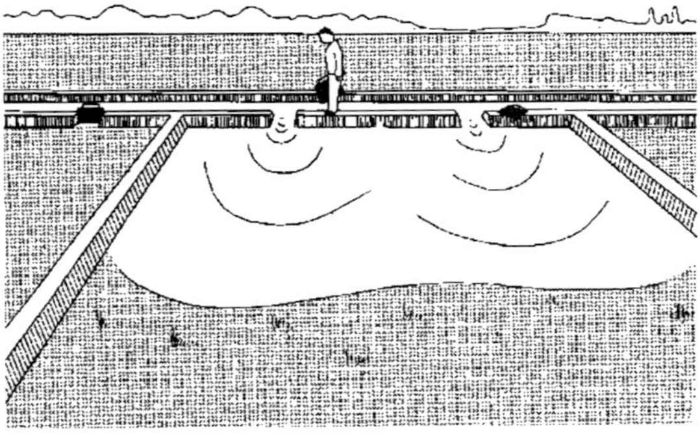

Fig. Border Irrigation

Basin Irrigation

Basins are horizontal, flat plots of land, surrounded by small dykes or bunds. The banks prevent the water from flowing to the surrounding fields. Basin irrigation is commonly used for rice grown on flat lands or in terraces on hillsides. Trees can also be grown in basins, where one tree usually is located in the centre of a small basin.

Sprinkler Irrigation

With sprinkler irrigation, artificial rainfall is created. The water is led to the field through a pipe system in which the water is under pressure. The spraying is accomplished by using several rotating sprinkler heads or spray nozzles or a single gun type sprinkler.

Drip Irrigation

In drip irrigation, also called trickle irrigation, the water is led to the field through a pipe system. On the field, next to the row of plants or trees, a tube is installed. At regular intervals, near the plants or trees, a hole is made in the tube and equipped with an emitter. The water is supplied slowly, drop by drop, to the plants through these emitters.

METHODS OF IRRIGATION

Water is an important constituent of the plant body. Plants need water to carry on different physiological processes essential to their growth and development. A great deal of water from plant body is wasted through the process of transpiration. Therefore, to compensate this loss of water, absorption of water from soil is an important phenomenon. Soil gets water mainly from two sources *viz.* precipitation and irrigation.

Hence, irrigation is a process, other than natural precipitation, which supplies water to crops, orchards, grass, or any other cultivated. plants. In the wetter parts of the world where rainfed cultivation is practiced, the farmer's activities consist of selecting suitable land, preparing the soil for cultivation; sowing, tending and harvesting the crops. Natural rainfall provides the water needed. But in many other places otherwise favourable for cultivation, natural rainfall does not provide all the water needed and, as such, irrigation of crops is required to make up this deficiency.

Techniques adopted for carrying water from its source to the crop are called methods or modes of application.

These are:

- Flooding -wetting all the land surface;
- Furrows -wetting only certain part of ground level;
- Sprinkler -in which the soil is wetted in much the same way as rains;
- Sub-surface irrigation -in which surface is wetted very lightly, but the sub soil is fully saturated; and
- Localized irrigation -in which water is applied at each individual plant at. a near daily rate.

CHARACTERISTICS OF AN EFFICIENT IRRIGATION METHOD

An efficient method of irrigation should fulfill five major objectives *viz.* (1) distribution of water uniformly over the field according to crop need, (2) storage of maximum fraction of water in the root zone for plant use, (3) crop growth should not be adversely affected, (4) soil transport or loss is negligible, and (5) the technique used is economically sound and adoptable at the farm.

FACTORS AFFECTING SUITABILITY OF IRRIGATION METHOD

The selection of a suitable irrigation method for a particular farm location depends upon the following factors.

Characteristics of Surface Soil

Textural, crusting, cracking and infiltration characteristics of surface soil; nature and depth of relatively impermeable layers in sub-soil, if any; water storage capacity of root zone; nature and extent of land slope; size of field;

surface drainage; nature and extent of salts in surface and sub-soil are the salient soil factors influencing between of an irrigation method.

Water Supply

Nature of water supply (continuous or rational). source (pump or canal), size of water delivery, quality of irrigation water, and quantity of water supply (adequate or limited) area few factors that must be taken into consideration while deciding the method of irrigation.

Nature of Crops

Nature of crops, area under different crops and their rooting behaviour, optimum depth and timing of irrigation, sensitivity of crops to excessive soil moisture, cultural operations required, etc. must be considered at the time of selection of irrigation method for a crop.

Others Factors of Irrigation

There are other factors influencing irrigation method like outlook, managerial efficiency and financial resources of the farmer; nature of the farm machinery used; availability and cost of labour; wear and tear maintenance facilities and costs of irrigation equipments; and availability of power supply. As far as possible, an irrigation method should not only provide a high level of water application efficiency, but also ensure its economic viability, sustained soil productivity and wide' adaptability to prevalent feature of the farm. Generally, irrigation methods followed in India lack in an economic use of irrigation water.

SPRINKLER METHOD OF IRRIGATION

In the sprinkler method of irrigation, water is applied above the ground surface as spray. The spray is developed by tlte flow of water under pressure through small orifices or nozzles. The pressure is obtained by pumping with careful selection of nozzle sizes, operating pressures and sprinkler spacing. High efficiency in water application/distribution can be obtained with sprinkler system.

Sprinkler systems are of generally two major types *viz.*

1 Rotating head system, and

2 Perforated pipe system.

In case of rotating head system small nozzles are placed on riser pipes and these riser pipes are fixed at an even interval along the length of lateral pipes which are placed on the ground surface. However, they can be mounted on posts exceeding the crop height and made rotating through 90 degree. In rotating sprinkler, the most, important device to rotate the sprinkler head is a small hammer activated by the trust of water striking the vane connected to it. In case of perforated pipe system, holes are perforated in lateral irrigation pipes which is especially designed to distribute water with a good deal of uniformity. This system is usually designed for low operating pressures (*i.e.* 0.5 to 2.5 kg/sq cm). Due to this low pressure, the system is attached to an overhead tank to achieve the requisite pressure head. The sprays are directed on both sides of the pipe which cover a strip of land from 6 to 15 metres wide. Nearly all cultivable soils can be sprinkler irrigated. It is, however, not suitable in very high textured soils where the infiltration rates are very low (*i.e.* less than 4 mm per hour). Most crops excepting rice and jute can be sprinkler irrigated. The flexibility of sprinkler equipment and efficient control of its application make this method adaptable to most of the topographic conditions. However, extremely high temperature and wind velocity markedly reduce the uniformity of water distribution and irrigation efficiency. This I system of irrigation is especially useful to the soils that have steep slopes or irregular topography and soils which are too shallow to level.

Advantages

- This technique enables judicious utilization of even small water flows and permits efficient irrigation of undulated lands, and soils with shallow depths.
- It saves 10 to 16 per cent land that is used in construction of channels and ridges in other methods.
- Highly permeable as well as relatively less permeable soils can be easily irrigated by sprinkler method without any risk of run-off and erosion, inundation and seepage losses.
- Fertilizers, pesticides and weedicides can be applied along with water spray, thus, saving extra labour.

Disadvantages

- High initial cost of equipments.
- Operating costs are generally higher than irrigation by surface methods.
- Winds disturb the sprinkler pattern giving uneven distribution of the irrigation water.
- Sprinkling with water containing an appreciable amount of salts may result in bum or death of the plants.
- Under certain climatic conditions diseases may be encouraged. The problem of fruit rotting in tomato and strawberry gets aggravated especially in moist soil condition.

LOCALIZED IRRIGATION

Drip Irrigation

As the name signifies, drip irrigation, also termed as trickle irrigation, involves the slow application of water to the root zone of a crop. The method was initiated in Israel and is now being tried in other countries. In this method, water can be used very economically, since loss due to deep percolation and surface evaporation are reduced to the minimum. This method, therefore, is highly suitable to arid regions and orchard crops. The successful raising of orchards even on saline soils has been made possible by the drip system of irrigation. The system can also be used for applying fertilizers in solutions.

The water is applied more frequently, close to the stems of plants through suitably spaced drippers (emitters) attached to plastic or metallic pipes spread above or below soils along crop rows. The pipes are hooked to source of water supply through a storage tank or pressure device which provides necessary hydraulic head or pressure for movement of water to the drippers. A pumping unit creates a pressure of about 2.5kg/sq cm. In this case only a part of soil in the vicinity of plant roots is wetted and kept close to field capacity. The amount of water dripping from nozzles can be regulated as desired by varying the pressure at the nozzles and the size of the orifice of the nozzles. The initial high cost of the equipment and its maintenance are the major limitations in this system. It may, however, work out to be cheaper than the sprinkler system especially for the orchards and other widely spaced crops.

Earthen Pot Irrigation Method

This method has been recently developed by the Haryana Agriculture University, Hissar. It is very cheap and convenient method and can be easily adopted. In this method, first a pit (60 cm deep and 90 cm wide) is dug out and the earthen pot is embedded up to the neck level and then it is filled with water. The water is absorbed by the soil through capillary action. One pot is sufficient to moisten one square meter area. Care should be taken to cover the pot by lid and supply water from time to time. The pots may be replaced by new ones after two or three seasons..

This system is especially useful for irrigating vegetable crops ego cucurbitaceous crops grown in sandy loam or sandy soils.

Advantages

- It is the most suitable method for vegetable and orchard crops where plants are widely spaced.
- It can be mly used in sandy and undulating lands.
- Saline water can be freely used because salts are deposited at the bottom of the pot
- High water use efficiency.
- A simple and comparatively cheap method.

Double Walled Pot "Jaltripti"

It is a useful irrigation device for the desert, developed by the Central Arid Zone Research Institute (CAZRI), Regional Station, Bikaner. It reduces the frequency and total amount of water needed and ensures a regulated constant supply of moisture to tree plants for survival and better growth. This device consists of a double walled earthen pot called "Jaltripti" (water satisfaction). The diameter of the outer pot is kept, approximately 25 cm at the top and at base it is 18 cm. The diameter; of inner pot is 15 cm on top and 12cm at base. The height is kept 30 cm. (Since it is prepared on potter's wheel, approximate dimensions have been given. Moreover, some variations in the measurement may occur). The dimensions of the inner pot have been kept slightly bigger than the size of the polythene bags (25 cm long measuring 10: cm across) used for raising plants in nursery. Both the pots are joined together at the base and the basal portion of inner pots is kept almost open. The external side of the outer pot is made impervious with the f help of paint, cement or coaltar.

The "Jaltripti" functions on two simple principles:

- Soil moisture tension and plant roots create a suction force which draws moisture towards it from the neighbouring high moisture zone.
- Earthen pots have many micro-pores in their wall which do not allow water to flow freely but allow its seepage in the direction where suction develops.

At the place where planting has to be done, the device is fixed in the soil in such a way that the brim of the outer pot is above soil surface. A tree sapling along with a soil ball is transplanted in the inner pot The water is filled in the space between two pots and the circular surface of water is caused by a polythene sheet to avoid direct evaporation of water. The paint on the external surface of the pot prevents outward movement of water through seepage. But the suction force created by the inner pot allows for the seepage of water steadily in that direction and keeps the soil sufficiently moist for the I growth of the plants. Water is filled weekly or fortnightly depending upon the season and size of the pot. The device has been named 'Jaltripti'.

Advantages

- Economy of water (80 to 90 per cent).
- It saves more water than drip and sprinkler system.
- Low irrigation frequency.
- Labour saving.
- No loss through percolation, evaporation and seepage.
- Regulated supply of moisture. Moisture is always available at field capacity.
- Promising in stabilization of sand dune..

Tueboponics of Field Irrigation

Tueboponics is a fairly recent development in the field of irrigation. This technique has been developed in Israel and practiced in the desert areas of that country in order to convert it into green lush forests. In tueboponics, water is provided to plants/trees through injections. The needle is inserted into the plant stem and water is delivered into the phloem. The needle and injection used for this purpose are of special shape and design prepared only for this purpose.

In India, this technique has not yet arrived but it may prove very useful in Indian deserves where water is very scarce. Its use in crop plants seems to be impracticable. It is a very simple device.

DRIP AND SPRINKLER IRRIGATION IN INDIA

India ranks first in respect of total irrigated area existing in the world. It has got approximately 80 million hectares of irrigated land. But the methods of irrigations employed are still very primitive and inefficient. Recent achievements in the field of irrigation for instance drip and sprinkler irrigations are yet not sufficiently popular in India. More than 10 million hectare is irrigated by sprinkler method and I million hectare by drip irrigation in the world. But in India, it is only about 0.7 m. ha under sprinkler irrigation and less than 20,000 ha with drip irrigation. Therefore, it is necessary to popularize these advanced methods of irrigation especially in those areas where water is a scarce resource.

India is blessed with abundant water resources. However, the available water, particularly for irrigation is tending to diminish and at the same time its demand is gravely felt due to population explosion. The emerging challenge is to tap all the available resources of water. Technological innovations are to be exploited to achieve the twin objectives of higher productivity and better water use efficiency.

For this, we will have to popularize drip and sprinkler irrigation methods. On account of certain financial, technical and institutional constraints, these methods have not got their due place in India and consequently, the area benefited is negligible. Therefore, the question arises as to what are the constraints and problems holding up progress.

The following are the major constraints faced by the farmers in adopting the drip and sprinkler systems of irrigation.

- High initial cost.
- Inadequate subsidy amount.
- Difficulty in getting subsidy amount
- Lack of availability of technical input and after sale services.
- Clogging of dripper and cracking of laterals.
- Damages due to rats and squirrels.
- High cost of spares and components.
- Discrimination in subsidy distribution among different categories of farmers.

To exploit the full potential of these two innovations, the constraints are to be overcome by appropriate policy instruments, financial support and technical guidance. This calls for an integrated approach and endeavour on the part of government. Implementing agencies, rnanufacturing cornpanies, voluntary organizations and the ultimate users of the systems *i.e.* the farmers.

METHOD OF IRRIGATING BASINS

There are two methods to supply irrigation water to basins:

1. The direct method and
2. The cascade method.

The Direct Method

Irrigation water is led directly from the field channel into the basin through siphons, spiles or bundbreaks.

The Cascade Method

On sloping land, where terraces are used, the irrigation water is supplied to the highest terrace, and then allowed to flow to a lower terrace and so on. The water is supplied to the highest terrace and is allowed to flow through terrace until the lowest terrace is filled.

This is a good method to use for paddy rice on clay soils where percolation and seepage losses are low. However, for other crops on sandy or loamy soils, percolation losses can be excessive while water is flowing through the upper terraces to irrigate the lower ones. This problem can be overcome by using the borrow-furrow as a small channel to take water to the lower terrace. The lower terrace is irrigated first and when complete the bund is closed and water is diverted into the next terrace. Thus the terrace nearest the supply channel is the last to be irrigated.

When long cascades are used for growing rice it is common practice to allow water to flow continuously into the terraces at low discharge rates. The water demand in the cascade can easily be monitored by observing the drainage flow. If there is no drainage then more water may be required at the top of the cascade. If there is a drainage flow then it is possible to reduce the inflow.

Wetting Patterns

For good crop growth it is very important that the right quantity of water is supplied to the root zone and that the root zone is wetted uniformly. If crops receive too little water, they will suffer from drought stress, and yield may be reduced. If they receive too much water, then water is lost through deep percolation and, especially on clay soils, permanent pools may form, causing the plants to drown. How much irrigation water should be supplied to the root zone - in other words "the net irrigation depth". How the irrigation water can be evenly distributed in the root zone is explained below, and an example of the evaluation of basin irrigation performance.

Ideal Wetting Outline

To obtain a uniformly wetted root zone, the surface of the basin must be level and the irrigation water must be applied quickly. An ideal wetting pattern: the basin is level and the right quantity of water has been supplied with the correct scream size. It is not possible to have the wetting pattern and root zone coincide completely. The part of the basin near the field channel is always in contact with the irrigation water longer than the opposite side of the basin. Therefore percolation losses will occur near the field channel, if sufficient water is supplied to the opposite side of the basin.

Poor Wetting Patterns

Poor wetting patterns can be caused by:

- Unfavourable natural conditions, *e.g.*, a compacted subsoil layer, or different soil types within one basin;
- Poor layout, *e.g.*, a poorly levelled surface;
- Poor management, *e.g.*, supplying incorrect stream size, applying too little or too much water.

Unfavourable Natural Conditions

A compacted sub-soil layer can sometimes occur in a basin some 30-50 cm below the soil surface. Infiltration through this layer may be very slow and so water tends to accumulate above this layer: a "perched" water table is formed. This may result in waterlogging.

This situation may be very helpful for growing rice but will be harmful for other crops. The compacted layer can be removed by using deep ploughs or rippers which break up the subsoil. Different soil types within a basin can cause very uneven water distribution. This problem can be solved by re-aligning basin boundaries so that each basin contains only one soil type.

Poor Layout

The wetting pattern if the soil surface is not level. Some parts of the root zone receive too little water and in the lower parts water may pond or be lost through deep percolation. Plants suffer in the drier parts because they receive too little water and wilt. Plants may also suffer in the wet parts; plant nutrients are carried away from the rootzone to the subsoil and, especially on clay soils the plants may drown. These faults can easily be corrected by careful land levelling.

Poor Management

The basin is irrigated too slowly, by using a stream size which is too small

The part of the basin which receives irrigation water first (near the supply channel) and thus the longest, receives too much water. Percolation losses occur, nutrients are washed away and the plants may drown. The other end of the basin remains too dry. The plants there do not receive enough water and wilt.

The solution to the problem is to:

- Increase the stream size so that the basin will be flooded more rapidly, or
- Subdivide the basin into smaller basins; smaller basins need a smaller stream size than larger basins.

If insufficient water is applied to fill the root zone. This is called "under-irrigation" and is caused by under-estimating the time needed to fill the root zone. There are no percolation losses during under-irrigation. Although water may be used efficiently by this approach, frequent irrigation will be necessary to meet crop water needs. However, continual under-irrigation will eventually restrict root development and the crop may suffer when there are delays in irrigating, *e.g.*, when water is in short supply or the supply system breaks down.

PRESERVATION OF BASINS

Bunds are susceptible to erosion which may be caused by, for example, rainfall, flooding or the passing of people when used as footpaths. Rats may dig holes in the sides of the bunds. It is therefore important to check the bunds regularly, notice defects and repair them instantly, before greater damage is done. Before each growing season, the basins should be checked to see that they remain level. During pre-irrigation it can easily be seen where higher and lower spots are; there should be smoothed out. Also, the field channels should be kept free from weeds and silt deposits.

PLANNING FOR WASTEWATER IRRIGATION

CENTRAL PLANNING

Government policy on effluent use in agriculture will have a deciding effect on what control measures can be achieved through careful selection of site and crops to be irrigated with treated effluent. A decision to make treated effluent available to farmers for unrestricted irrigation or to irrigate public parks and urban green areas with effluent will remove the possibility of taking advantage of careful selection of sites, irrigation techniques and crops in limiting the health risks and minimizing environmental impacts.

However, if a Government decides that effluent irrigation will only be applied in specific controlled areas, even if crop selection is not limited (that is, unrestricted irrigation is allowed within these areas), public access to the irrigated areas will be prevented. Without doubt, the greatest security against health risk and adverse environmental impact will be achieved by limiting effluent use to restricted irrigation on controlled areas to which the public has no access but even imposing restrictions on effluent irrigation by farmers, if properly enforced, can achieve a degree of control.

Cobham and Johnson have suggested that the procedures involved in preparing plans for effluent irrigation schemes are similar to those used in most forms of resource planning and summarized the main physical, social and economic dimensions as in Figure. They also indicated that a number of key issues or tasks were likely to have a significant effect on the ultimate success of effluent irrigation, as follows:

- Organizational and managerial provisions made to administer the resource, to select the effluent use plan and to implement it,
- The importance attached to public health considerations and the levels of risk taken,
- The choice of single-use or multiple-use strategies,
- The criteria adopted in evaluating alternative reuse proposals,
- The level of appreciation of the scope for establishing a forest resource.

Adopting a mix of effluent use strategies is normally advantageous in respect of allowing greater flexibility, increased financial security and more efficient use of the wastewater throughout the year, whereas a single-use strategy will give rise to seasonal surpluses of effluent for unproductive disposal. Therefore, in site and crop selection the desirability of providing areas for different crops and forestry so as to utilize the effluent at maximum efficiency over the whole yearly cycle of seasons must be kept in mind.

DESIRABLE SITE CHARACTERISTICS

The features which are critical in deciding the viability of a land disposal project are the location of available land and public attitudes. Land which is far distant from the sewage treatment plant will incur high costs for transporting treated effluent to site and will generally not be suitable. Hence, the availability of land for effluent irrigation should be considered when sewerage is being planned and sewage treatment plants should be strategically located in relation to suitable agricultural sites. Ideally, these sites should not be close to residential areas but even remote land might not be acceptable to the public if the social, cultural or religious attitudes are opposed to the practice of wastewater irrigation.

The potential health hazards associated with effluent irrigation can make this a very sensitive issue and public concern will only be mollified by the application of strict control measures. In arid areas, the importance of agricultural use of treated effluent makes it advisable to be as systematic as possible in planning, developing and managing effluent irrigation projects and the public must be kept informed at all stages. The ideal objective in site selection is to find a suitable area where long-term application of treated effluent will be feasible without adverse environmental or public health impacts. It might be possible in a particular instance to identify several potential sites within reasonable distance of the sewered community and the problem will be to select the most suitable area or areas, taking all relevant factors into account. The following basic information on an area under consideration will be of value, if available:

- A topographic map,
- Agricultural soils surveys,
- Aerial photographs,
- Geological maps and reports,
- Groundwater reports and well logs,
- Boring logs and soil test results,
- Other soil and peizometric data.

At this preliminary stage of investigation it should be possible to assess the potential impact of treated effluent application on any usable aquifer in the area(s) concerned. The first ranking of sites should take into account other factors, such as the cost and location of the land, its present use and availability, and social factors, in addition to soil and groundwater conditions.

The characteristics of the soil profile underlying a particular site are very important in deciding on its suitability for effluent irrigation and the methods of application to be employed. Among the soil properties important from the point of view of wastewater application and agricultural production are: physical parameters (such as texture, grading, liquid and plastic limits, etc.), permeability, water-holding capacity, pH, salinity and chemical composition. Preliminary observation of sites, which could include shallow hand-auger borings and identification of vegetation, will often allow the elimination of clearly unsatisfactory sites.

After elimination of marginal sites, each site under serious consideration must be investigated by on-site borings to ascertain the soil profile, soil characteristics and location of the water table. Peizometers should be located in each borehole and these can be used for subsequent groundwater sampling. A procedure for such site assessment has been described by Hall and Thompson and, if applied, should not only allow the most suitable site among several possible to be selected but permit the impact of effluent irrigation at the chosen site to be modelled. When a site is developed, a long-term groundwater monitoring programme should be an essential feature of its majagement.

CROP SELECTION ISSUES

Normally, in choosing crops, a farmer is influenced by economics, climate, soil and water characteristics, management skill, labour and equipment available and tradition. The degree to which the use of treated effluent influences crop selection will depend on Government policy on effluent irrigation, the goals of the user and the effluent quality. Government policy will have the objectives of minimizing the health risk and influencing the type of productivity associated with effluent irrigation. Regulations must be realistic and achievable in the context of national and local environmental conditions and traditions.

At the same time, planners of effluent irrigation schemes must attempt to achieve maximum productivity and water conservation through the choice of crops and effluent application systems. A multiple-use strategy approach will require the evaluation of viable combinations of the cropping options possible on the land available. This will entail a considerable amount of survey and resource budgeting work, in addition to the necessary soil and water quality assessments.

The annual, monthly and daily water demands of the crops, using the most appropriate irrigation techniques, have to be determined. Domestic consumption, local production and imports of the various crops must be assessed so that the economic potential of effluent irrigation of the various crop combinations can be estimated.

Finally, the crop irrigation demands must be matched with the available effluent so as to achieve optimum physical and financial utilization throughout the year. This process of assessment is reviewed by Cobham and Johnson (1988) for the case of effluent use in Kuwait, where afforestation for commercial purposes was found to offer significant potential in multiple-use effluent irrigation.

CONVENTIONAL IRRIGATION SYSTEMS AND WATER UTILIZATION

The present day conventional irrigation systems have major disadvantages. The distribution and drainage at the farm level are left to the individual farmers. The present irrigation practices are often extremely wasteful. Although 3 the prevailing practices vary widely, estimates indicate that farmers lose more than half the amount of water they receive at the headgate, through evaporation, runoff, or deep percolation.

CONSTRAINTS OF WATER UTILIZATION

The following are some of the major constraints in the proper water utilization. In gravitational irrigation water can not be used uniformly and efficiently by the farmers, if the land is not properly levelled. Often the farmers do not have the adequate knowledge about land levelling and shaping. For better irrigational efficiency, land levelling is a prerequisite though it is expensive also. The irrigational projects usually have a definite command area, which can be brought under irrigation. The main activities under these projects are on farm works such as construction of field channels and drains, promoting small bunds across small streams, organizing supply facilities etc. Canal water is easily and conveniently available to the fields which are situated at low level and at the head ends of the channels while it is scarcely available at the tail ends and higher fields. Wherever the lift irrigation from canal is permitted, it becomes very costly and uncertain.

Due to construction of 'Kutcha' channels a large quantity of water is lost in transporp due to percolation, leakages etc. Thus overall water requirement increases many more times than actually required for physiological processes of crop growth. Sometimes accumulation of water in the field is also observed as waste due to lack of proper drainage.

In the conventional irrigation system, from outlet (distributions) to the fields, channels are made in the field which usually are subjected to leakages etc. This requires constant supervision by labourers so that irrigation water is not lost to other fields. For distribution and irrigation also labourers are required.

Disadvantage of Conventional Irrigation

The important disadvantages of the irrigation methods conventionally followed in India.

- An estimated level of 43 per cent efficiency in surface water utilization and 70 per cent efficiency in ground water utilization (at farm level) recorded, shows that a large quantity of water is not properly utilised in the present irrigation system.
- The distribution of water in the field by main canal, sub-canals and distributions up to the outlet points into the fields and subsequently in the far off drainage points in the fields, is uneven due to poor levelling. This sometimes creates problem for getting sufficient and adequate irrigation to the crops.
- Crops are usually subjected to cyclic changes of flooding and water stress situations, by providing heavy irrigation at one time and leaving the fields to dry up for about 10 to 15 days. The moisture availability to the crops is fluctuates from saturation to stress and again to saturation. This results in poor yields of the crops.
- The fields situated in low areas always get excess water causing prolonged water logging due to lack of levelling of fields. Thus crops are subjected to water logging resulting in poor yields.
- In the fields about 10-15 per cent of land is utilized for preparing channels and distributions etc. which decreases effective area of cultivation.
- Extensive areas of land in the arid and semi-arid regions of India have gone out of cultivation due to rise of water table and accumulation of salts. Excessive irrigation and poor water management are the chief reasons of water logging and gradual

build up of excessive salts. Progressive build up of soil salinity has made the soils unsuitable for cultivation.

- Sometimes a large quantity of irrigation water is subjected to deep percolation and seepage resulting in the use of higher amount of fertilizers.
- In the absence of planned proper combinations of spacing, length and slope of furrow's and suitable size of irrigation streams and duration of the water application, it will be very difficult in regulating the flow of irrigation water in the fields.
- Slow and gradual changes in soil physical properties, particularly soil structures due to top soil erosion, takes place under present irrigation practices.
- Cyclic changes of wetting and drying of soil particles, around the roots result in disturbances of root activity of the crop.
- In conventional irrigation, unmanageable undesirable weed growth is usually observed. These weeds hinder free movement of water.
- Due to the improper irrigation methods in the conventional irrigation system, crops can not be raised satisfactorily. The moisture in the soil fluctuates between too dry to too wet stages. The moisture fluctuation in the conventional system of flood irrigation.

Unsurmountable Problems

In the conventional irrigation methods, a large number of problems are faced. Some of them are briefly explained here. The canal and lift irrigation system is so designed that farmers cannot have choice of crops. Generally in the command area all the farmers are compelled to chose the same crop and even the same varieties is the water is available only for certain periods. Besides, due to seepage and percolation of water on both sides of the canal water logging, salinity and alkalinity are common in the farmers fields. Since wells are quite deep, the costs for digging or boring, high lifting pumps, their installment and maintenance on electricity or diesel have to be borne by the farmers.

The intensity of problems due to flow irrigation method varies from field to field depending upon the soil profile and topography of the soils. Too clayey and sandy soils create hurdles in irrigation as the former offers much resistance to penetration of water and in the later water is easily lost through deep percolation and seepage.

DIVERSION STRUCTURES IN IRRIGATION SYSTEMS

Most surface irrigation systems derive their water supplies from canal systems operated by public or semi-public irrigation departments, districts, or companies. Some irrigation water is supplied in piped delivery systems and some directly pumped from groundwater.

Diversion structures perform several tasks including.

- On-off water control which allows the supply agency to allocate its supply and protects the fields below the diversion from untimely flooding;
- Regulation and stabilization of the discharge to the requirements of field channels and watercourse distribution systems;
- Measurement of flow at the turnout in order to establish and protect water entitlements; and
- Protection of downstream structures by controlling sediments and debris as well as dissipating excess kinetic energy in the flow.

Conveyance, Distribution and Management Structures

Conveying water to the field requires similar structures to those found in major canal networks. The conveyance itself can be an earthen ditch or lateral, a buried pipe, or a lined ditch. Lined sections can be elevated or constructed at surface level. Pipe materials are usually plastic, steel, concrete, clay, or asbestos cement, or they may be as simple as a wooden or bamboo construction. Lining materials include slip-form cast-in-place, or prefabricated concrete, shotcrete or gunite, asphalt, surface and buried plastic or rubber membranes, and compacted earth.

The management of water in the field channels involves flow measurement, sediment and debris removal, divisions, checks, drop-energy dissipators, and water level regulators. Some of the more common flow control structures for open channels. Associated with these are various flow measuring devices like weirs, flumes, and orifices. The designs of these structures have been standardized since they are small in size and capacity. Designs for flow measurement and drop-energy dissipator structures need more attention and construction must be more precise since their hydraulic responses are quite sensitive to their dimensions.

Field Distribution Systems of Irrigation

After the water reaches the field ready to be irrigated, it is distributed onto the field by a variety of means, both simple and elaborately constructed. Most fields have a head ditch or pipeline running along the upper side of the field from which the flow is distributed onto the field.

In a field irrigated from a head ditch, the spreading of water over the field depends somewhat on the method of surface irrigation. For borders and basins, open or piped cutlets are generally used. Furrow systems use outlets which can be directed to each furrow.

A system in which siphon tubes are used as a means of serving each furrow. Field distribution and spreading can also be through portable pipelines running along the surfaces or permanent pipelines running underground. Basins and borders usually receive water through buried pipes serving one or more gated risers within each basin or border. A typical riser outlet, known as an alfalfa valve. The most common piped method of furrow irrigation uses plastic or aluminium gated pipe. The gated pipe may be connected to the main water supply via a piped distribution network with a riser assembly like, directly to a canal turnout, or through an open channel to a piped transition.

DRIP IRRIGATION - INDIA

Drip irrigation is of recent origin, and, in India, is being used on a limited scale in Tamil Nadu, Karnataka, Kerala and Maharashtra States, mainly for coconut, coffee, grape and vegetable production. Drip irrigation systems (DIS) are extremely effective in arid and drought prone areas where water is scarce, and have been used experimentally in India for over 15 years: in the States of Tamil Nadu, Karnataka, Maharashtra, and Andhra Pradesh, progressive farmers started using this method of irrigation in the late-1970s without the benefit of any subsidies or support from central or state governments. However, as a result of subsequent, sustained efforts by the state and central governments, agricultural universities, and private sector manufacturers, use of drip irrigation systems spread through the drought prone areas of southern and western India. The use of DIS, however, is primarily to irrigate high value, horticultural crops. In states like Maharashtra, Karnataka, and Tamil Nadu, DIS are sometimes used for irrigation of vegetable and other commercial crops. The sharp rise in the area under DIS irrigation between 1988 and 1989 is due, in large part, to the significant increase in the use of these systems in the Maharashtra State.

Technical Description

Drip irrigation systems deliver water and agrochemicals (*e.g.*, fertilizers and pesticides) directly to the root zones of the irrigated plants at a rate best suited to meet the needs of the plants being irrigated. Thus, this system makes efficient use of water, especially when compared to conventional methods of irrigation such as furrow, border, basin and sprinkler irrigation systems, which, under arid and drought conditions, suffer from an high rate of water loss and have a low degree of water use efficiency.

Operation and Maintenance

The principle operation and maintenance requirements associated with the implementation of this technology include the need for regular cleaning of the system and careful monitoring of the quality of the source water, as the drip irrigation systems are very sensitive to the clogging of the drippers. The systems also require a relatively high degree of skill to design, install and operate, and are susceptible to theft, damage and disruption by rodents that destroy the drip pipes and drippers.

Level of Involvement

The use of this technology requires skilled personnel. Because of the relatively high capital cost of the piping systems necessary to implement this technology, the initial funding for the project may require some level of government involvement. Regular operation and maintenance of the system is the responsibility of the individual operator.

Costs

The capital costs involved in the establishment of a drip irrigation system are high compared to the costs of establishing conventional irrigation systems. However, the labour requirements and operational costs are low. The net result is that the benefit-cost ratio for DIS is very favourable compared to conventional systems since the payback period for investment very short. In the case of the orchard crops in Maharashtra, the cost of DIS ranged from $450/ha to $1 150/ha in 1990. Elsewhere, the cost of using drip irrigation systems for sugarcane irrigation averaged $715/ha, for banana irrigation $1 150/ha, and for cotcrus-fruit irrigation $575/ha, with the payback periods ranging from 2 months for banana crops, 12 months for cotcrus-fruit crops, and 18 months for surgarcane crops. Comparative benefit-cost ratios for various crops ranged from 1.64 for groundnuts (peanuts), to 4.84 for pomegranates, to 5.15 for tomatoes, to 8.58 for grapes, to 15.0 for mosambi. These ratios compare to benefit-cost ratios of 1.80, 2.20, 3.96, 6.38, and 9.81, respectively, using conventional irrigation systems.

Effectiveness of the Technology

In almost all of the cases reported, excepting the Jobner case, there was an improvement in crop yields and savings in water use of between 18 per cent and 40 per cent. Consequently, there was a substantial improvement in the water use efficiency that ranged up to three times that of water use efficiencies achieved using conventional surface irrigation methods, even with the use of poor quality irrigation water. Because of the directed delivery of irrigation water, it is possible to utilize poor quality irrigation water using the drip irrigation system. The performance of this technology is summarized in Tables. The data presented in Table are based upon water savings and increased yields achieved in Maharashtra State using drip irrigation systems. In addition to the improved yields and water savings, for crops such as sugarcane there is a savings in labour costs that equals the savings in water.

Advantages

The advantages of drip irrigation systems include an high efficiency of water use and greater crop yields compared to other irrigation methods. In addition, crops irrigated using drip irrigation systems generally require less tillage and are of better quality. DIS also contribute to improved plant protection and reduced occurrences of plant diseases and greater efficiencies in the use of fertilizers, because water containing the agrochemicals is applied directly to the plant roots in the quantities necessary for optimal plant production. For a similar reason, DIS can also make use of lower quality water, and results in no return flows, tail water losses or increased soil erosion. Because water is applied in optimal quantities, plants generally have a shorter growing season and produce fruit earlier, with less weed growth and pest damage than conventionally irrigated crops.

The lower labour requirements result in relatively low operational costs, with savings in labour of up to 90 per cent of the costs associated with conventional systems, in part, because mechanical operations can be carried out simultaneously with the application of irrigation water. DIS can be used in hilly terrain and on lands with problem soils, and results in improved infiltration in soils with low conductivity. Drip irrigation systems are low pressure systems, which can be adapted for use in greenhouses, and with automated control systems.

Disadvantages

Drip irrigations systems have a sensitivity to the clogging of the drippers, which may require pretreatment of turbid source waters, and, if not properly installed, can cause moisture distribution problems. The systems are also susceptible to rodent damage. The systems have an high cost compared to conventional irrigation methods, and require higher levels of skill for design, installation, and operation, which make them liable to damage or theft.

USES OF IRRIGATION WATER

In most of the above uses, irrigation water has multiple effects. Some of the irrigation induced impacts are desirable and beneficial to society while others are undesired and adverse in terms of their negative impacts on humans and environment (ecosystems). Some of the impacts are significant, while others may be insignificant—some of them are known while others remain unknown. Some of the impacts are common to most irrigation development projects, while others are more specific to certain locations, schemes or methods of irrigation. These impacts may be broadly classified into desired impacts (benefits) and undesired or negative impacts (costs). In a project level comprehensive assessment of irrigation impacts, following major impacts, both positive and negative impacts, can be identified.

Major Positive Impacts

(I) direct positive impacts of irrigation include:

- Increased agricultural production
  - Increased crop productivity
  - Expansion in crop areas

  - Increase in cropping intensity
  - Increase in crop diversification
- Increased commercial fish production (in-land fisheries)
- Increased benefits of water use in industrial, commercial and residential sectors-from raw water provided through irrigation infrastructure or from groundwater
- Increased environmental benefits of water for in-stream flows, disposal of waste, wildlife, flora and fauna; increased farm forestry and vegetation in irrigated areas.
- Increased health benefits-improved sanitation due to better access to water.
- Other direct positive impacts
  - Increased benefits from flood control
  - Increased benefits from water use for rural domestic and livestock purposes
  - Increased groundwater recharge; reduction in opportunity costs of water uses
  - Increased recreation from water bodies, sight seeing, fishing

Secondary Impacts of Irrigation

- Increased employment in agriculture due to increased cropping intensity, increased crop area and output from irrigation
- Increased employment outside agriculture from increased crop output in related industries such as input industry (backward linkages) and output processing industries (forward linkages)
- Positive impact on poverty reduction through increased productivity and increased employment opportunities
- Increased food security at national, regional and local levels
- Lower food prices for consumers, due to productivity gains and increased overall food supplies
- Improved nutrition, improved calorie intake and improved health

Major Negative Impacts

Irrigation may have several potential negative impacts on agro-,eco- and human systems.

These impacts may be also broadly divided into three categories:

1. *Adverse economic impacts:* Higher subsidy, distorted market and a relative neglect of rainfed farming and other less favoured farming practices
2. *Adverse social impacts:* Forced displacement and involuntarily resettlement of the local inhabitants
3. *Adverse environmental impacts:* Loss of biodiversity, obstruction on natural hydrological flows and damages to aquatic ecosystems

Some of the impacts are direct and enter into normal market process, whereas other negative impacts may be indirect and their costs may not be accounted for by individual decision-makers, nor are these costs are captured into the normal marketing activities— they are called as "externality effects." The term "externality" is generally defined as an effect or outcome when production or consumption of one party affects the production or consumption of another party and neither party makes any compensation for the effect. For example, costs imposed on downstream users of irrigated water polluted by upstream users. These are costs imposed on a society as a result of an action—irrigation development in this case. Based on the nature of the impacts on the surrounding environment and the changes they will bring in the hydrology and in the surrounding regions, environmental impacts or externalities generated by irrigation can be broadly grouped into three categories.

They are:

- *Change in water quantity:* This is related to where water goes and how it is used, *e.g.*, localized scarcity, water stress, etc.
- *Change in water quality:* This is related to change in quality and usability of water, *e.g.*, water contamination, salt loading, etc.
- *Change in soil quality:* This is related to impacts of water uses on land fertility and changes in land quality, *e.g.*, salinity build up, impacts on soil quality by changes in the water table, etc.

Some of the potential negative externalities associated with irrigation may include the following:

- Irrigation-induced land degradation
  - Soil salinity and water logging-on-farm and off-farm impacts
  - Loss of soil fertility due to irrigation induced crop intensification
  - Increase in biological imbalances due to irrigation (weeds, pests)
- Surface water pollution-nutrients/chemicals
- Groundwater pollution-nutrients/chemicals
- Toxic concentration of substances in surface and groundwater-salts, metals and pesticides
- Saline return flows
- Health impacts in terms of increased water borne diseases (schistosomiasis, malaria)
- Loss of bio-diversity-birds, fish and other wildlife species extinction
- Negative impacts on wetlands
- Social impacts
  - Communities displaced by large scale irrigation development
  - Loss of cultural heritage and historical places
  - Displacement of unskilled labour in mechanized irrigation

Given the wide range of these impacts, it may be useful to identify the causal factors for these impacts *i.e.* whether the impacts are due to farm level agricultural practices, field level water application systems, water distribution systems, water supply or drainage systems or large reservoirs or tanks.

THE PRESENT AND FUTURE SCOPE OF IRRIGATION

The area of land irrigated in the world nowadays is close to 250 million hectares, of which about two-thirds is situated in five countries-China, India, Pakistan, the Soviet Union and the USA. With world populations growing concurrently with expectations of higher standards of living, world food and fibre demands will increase. Since arid lands become highly productive with the introduction of irrigation water, and the productivity of rainfed areas increases considerably with supplementary irrigation, the improvement of existing irrigation projects and the opening up of new irrigated areas will continue to make significant contributions to solving the world's expanding food and fibre needs. In order to ensure the stability and permanency of these projects we cannot afford to overlook the potential hazards intrinsic to irrigation development, and we must be aware of the land transformations that may take place.

IRRIGATION DEVELOPMENT OF LAND TRANSFORMATIONS

Land transformations resulting from irrigation development are in all cases spectacular-especially when seen from the air or recorded in satellite imagery. Aerial photographs showing the boundaries between the desert and the sown in, for example, the Nile Valley in Egypt, the Central Valley in California, or the Rajasthan Canal in India, exhibit dramatic pictures of the contrasting land surfaces and highlight the transformations that have taken place as a result of irrigation making the deserts blossom.

These land transformations can be classified into two categories:

1. Surface modifications resulting from implementing various irrigation methods;
2. In-depth changes caused by disturbing the local or regional salt and water balance.

Most of these transformations are irreversible and permanent. All irrigation projects involve transferring water from one area to another-from a water- surplus region to a water-deficient one. Some major irrigation schemes call for inter-regional water transfers from one basin to another using major dams for inter-annual or inter-seasonal storage. They often require main canals bundreds of kilometres in length to transport the water. Inherent in this irrigation development are the land transformations caused by the construction of these large multi-purpose dams-such as those that have followed the building of the Aswan, Kariba, Tarbella and Boulder Dams. Reservoir sedimentation, steam bed aggradation, flooding of upstream forests and villages are only some of the many problems concomitant to major schemes. This chapter, however, is confined to land transformations that are directly related to the irrigation development.

METHOD OF SURFACE TRANSFORMATIONS

Irrigation water is applied to land using one of the four main methods:

1. Surface flooding;
2. Partial wetting of the surface in furrows;
3. Surface sprinkling;
4. By drip (or 'trickle').

All methods, particularly surface ones, involve considerable land preparation. The first land transformation follows clearing the land of trees, brush or other vegetation. In the semi-arid zones (for example in the western part of the United States) relatively dense cover of sagebrush, grease-wood, chapar- ral, scrub oak, pine or mesquite have to be eradicated from proposed irrigated lands, and in the monsoon areas such as Thailand, India, Indonesia, etc., tropical and sub-tropical jungle forests must be cut down and eliminated before development can begin. Small brush up to 0.5 m in height can be removed by ploughing, while vigourous vegetative growth usually requires the clearing operations to be carried out with large and powerful tractors. In some cases deep-rooted tree stumps must be blasted away-leaving the land with a bomb-pitted look of a war zone. In other cases stones, boulders and rock outcrops have to be hauled away from the project lands.

The land having been unoccupied, it next becomes necessary to smooth (or grade the surface) to facilitate uniform water distribution. These works bring about another land transformation. Surface smoothing is usually accomplished by the use of harrows, graders, scrapers, bulldozers and land planes- but in under-developed countries hand labour and bullock-drawn wooden planning boards are often used for this work. On accurately leveled land, water will be distributed uniformly over the fields; but when the leveling is done poorly and the ground surface remains rough and irregular, some parts of the fields will receive more water than others. The final shape of the land surface will depend on the method of irrigation to be used. In many of the older irrigation systems, where the wild-flooding method is common, water flows from the supply ditches over the fields and is guided only by the slope of the land.

This method requires smallest amounts of land leveling (say, up to about 200 cubic meters of earth moving per hectare), but irrigation by this method is generally the least efficient of all methods. More efficient methods of surface flooding call for accurate land shaping and grading and 1500 cubic meters of earth may seen to be moved per hectare. In the border method of irrigation the land is divided into long, narrow strips 5-15 m wide, extending for 100-400 m down the prevailing land slope, and the border strips are separated by levees, 0.1-0.3 m high. Border strips are leveled transversely, so that water will flow at an even depth per the whole width. The longitudinal slope is graded to a design gradient to enable an even percolation of the irrigation water over the entire border. In the level basin method the land is shaped into accurately leveled plots ranging in size from 0.2 to 1.0 ha and surrounded by low bunds. The basin technique is suited both to permeable soils which must be covered quickly with water in order to prevent excessive losses near the supply ditches, and to heavy soils with slow percolation rates. It is the method most commonly used for rice ('paddy') irrigation all over the world. Small, level basins are also used for irrigating orchards. Leveling works for constructing either borders or basins transform the original natural land surface into clearly discernable rectangular or square patterned fields. In the irrigation methods described above, almost the entire land surface is wetted at the time of irrigation. When furrow irrigation is used, the fields are only partially wetted. Furrows (20-30 cm in depth) - or corrugations (10-15 cm) are spaced according to the plant rows-one furrow or corrugation being provided for each row. For efficient furrow irrigation the land must be graded accurately and the flow of water into the furrows controlled.

The pattern created by this method is neither square or rectangular, but the fields are formed into long, thin, parallel linear strips leading away from the delivery ditches. In sprinkler irrigation projects, water is practical to the crop in the form of a spray. Unlike the surface methods, sprinkler irrigation requires less land leveling and can be used economically on undulating lands, on soils too shallow to prevent proper leveling, and on soils too porous for good water distribution by surface methods. The land transformation process is partly dependent on whether the sprinkler system is semi-permanent or portable. A particularly dramatic land metamorphosis takes place when central pivot installations are introduced on a large scale. In these schemes, combinations of rotating and fixed sprinklers deliver water from central pivots through continually rotating booms to circular-shaped fields.

The fields can have diameters of up to 1 km. These systems when seen from the air or on satellite photographs show up as green circles in a patterned design on a beige desert background. Drip irrigation systems consist of extensive networks of small-diameter plastic pipes delivering low flows of a few litres per hour to the plants through special emitters. From each emitter, water spreads laterally and vertically in the soil profile. Drip systems require little land leveling. Dissolved salts are left in the soil near the edge of the wetted zone and these salts must be periodically leached out of the root zone by rainfall or by irrigation.

The system is adaptable to a wide variety of shallow, sandy or stoney soils as well as undulating and steep topographic conditions. With the world's expanding population pressures, planners will be forced to bring the more difficult areas into production, and we can expect that extensive areas will come under drip irrigation in the future to enable otherwise unirrigable areas to be developed.

2

Function of Irrigation Management

Local associations of water users serve the same functions as irrigation agencies, but on a very localized level. In some countries such as Nepal and the Philippines, the major portion of the irrigated area is managed locally, through village-based water user associations. Typically such associations manage very small irrigation canals that were constructed by the users, perhaps centuries ago and the associations have grown up around the need for operating and maintaining these canals. In many developing countries such traditional canal systems have been the target of modernisation efforts by government agencies, often funded through international development assistance. The relatively crude physical infrastructure of many of these canal systems was rebuilt and absorbed into the domain of the government agency, which replaced the management functions of the indigenous water user associations.

Today, local water user associations are recognised as resources of 'social capital' which will have an increasingly important role to play in the coming decades. Instead of absorbing such associations into the state organisation, the capacity of associations is developed so that they can improve the performance of their own irrigation and drainage systems. This participatory approach was re-invented in the Philippines in the 1970s and has since evolved into a global trend that combines participation with various degrees of privatisation. In many developing countries a major focus of this process is the institutional challenge of establishing new water user associations which can serve the management functions previously handled by government. A particularly fruitful source for a case study of this trend is Mexico. With 5 million irrigated hectares it was the seventh largest irrigator in the world in 1995. It began the irrigation management transfer process as early as 1989 and the experience has been extensively researched. Finally, Mexico has for a number of years been considered as a paradigm for this form of institutional transformation. The case study draws heavily on the work of Kloezen, Garcés Restrepo and Johnson. The broad political and economic context within which Mexico introduced irrigation management transfer was central government's neoliberal response to the economic crisis of the 1980s and the unsatisfactory performance of the irrigation districts under the management of the National Water Commission. Transfer has been a top-down process motivated by the international development banks, the main objective being to reduce public expenditure on irrigation OM&M whilst promoting greater user participation in irrigation management.

It has been accompanied by a new National Water Act, revision of the Constitution to give legal foundation for the privatization of the *ejidos* (land reform communities) in all irrigation districts and termination of guaranteed crop prices and subsidized credit. 'Dismantling the public sector, including the public irrigation sector, would not have been possible without the commitment at the highest political levels to reduce staff working in the public sector'. The institutional transformation has been rapid and radical. In its first phase the National Water Commission retained management of the headworks and the main canals. Water user associations took over financial and managerial responsibility for OM&M below the main canal. In the second phase the Commission is left with management of the reservoirs and surface-water-pumping stations; a District Federation of the user associations takes on the OM&M of the main canals. These associations are non-governmental local entities for collective action.

The high speed of the process and the low resistance of farmers have surprised many observers. It seems that transfer was preceded by other neoliberal reforms in agriculture. Farmers knew the Commission's traditional OM&M was to be terminated and OM&M had in any case declined in quality. Reform came first in the larger districts in the north, where many large private producers were known to support change. Lastly, the programme built on a strong organizational base-the *ejidos* themselves and the private farmers' co-operatives and unions.

Government informed farmers of the impending change-there seems to have been no serious consultation-and instructed them to select their delegates to the new user associations. Thereafter, delegates to each association elected from among themselves a president, a treasurer and a secretary. The Commission worked with the new user groups for 6 months or more from the time of transfer and provided their leaders and technical staff with extensive training in OM&M and financial management. The Commission also granted concessions to the associations to use its machinery and equipment so that there would be no capital expenditure required in advance.

Prior to transfer the National Water Commission was wholly responsible at the district level for the planning of annual and seasonal water allocations. Similarly the Commission employed the heads of the irrigation units into which each district was divided. These units were more or less independent hydraulic blocks with sizes ranging from 3,000 to 20,000 ha. Unit heads and their channel-keepers were responsible for daily OM&M at all system levels down to the farm inlets. Farmers paid their fees at the unit office and were given water by the channel-keeper. After transfer, hydraulic committees were introduced at district level for allocation planning; these committees were composed of representatives from the Commission, the state government and each user association in that district. At the same time the district 'units' were replaced by 'modules'-two or more per unit-and it is at this level that the new water user associations manage. They collect irrigation water fees directly from farmers and employ their own module managers, channel-keepers, maintenance personnel and administrative staff. Based on the volume of water a module takes from the Commission and on the proportional amount of main infrastructure serving a module, each association must pay a percentage of the total fees it collects to the Commission.

The specific subject of this case study is the Alto Río Lerma Irrigation District. Located in the State of Guanajuato, it has an area of 113,000 ha. There are some 24,000 farmers of whom 55 per cent are from the *ejidos* and 45 per cent are private growers. The average landholding is 5 ha. Private holdings average twice the size of *ejido* holdings. The district's climate is sub-humid with annual precipitation of 750mm and potential evapotranspiration of 1900mm. Eighty millimetres of rain falls in the winter season from November to April and 670 mm falls from May to November. Average temperature is 19 °C and relative humidity is 60 per cent.

The Alto Río Lerma Irrigation District enjoys access to both surface water and groundwater. The river has four storage dams with a combined capacity of 2,140 mcm, as well as five diversion dams. The distribution network has 475 km of main canals and 1,660 km of secondary and tertiary canals, as well as 1,030 km of drainage canals. There are 1,710 deep wells exploiting three different aquifers with a total annual recharge of 500 mcm. Seventy per cent of the irrigated hectarage uses surface water and 30 per cent uses groundwater. Wheat and barley are the main crops in the dry winter months. Sorghum, maize and beans predominate in the wet summer season. All farmers grow vegetables and the private growers do so for the export market.

At the start of each agricultural year in November the hydraulic committee decides on the total area that can be safely irrigated in the district and by each of the eleven modules. This total area is derived from the combined volume of water in the four storage dams. The committee next sets the number of times that irrigation services can be delivered to each farmer in each season, typically 3-5 times in the winter and once in the summer. Each module receives its allocated share of surface water over the year and, given the module's ability to restrict the area irrigated by users, farmers can request water at any time within the constraints of the seasonal maximum. Farmers pay a fee to their water user association prior to each individual irrigation and get a receipt that has to be shown to the channel-keeper before water is allocated to their fields.

Kloezen and his colleagues show a keen awareness of the political and macroeconomic context of irrigation management transfer as well as a familiarity, for example, with Fox's sociological work on building social capital from below. However, the language of common property resources, freeriding and transaction costs is absent from their reports. What we do learn is that the number of staff engaged in governance, OM&M, administration, monitoring and evaluation *rose* by 13 per cent between the 'before' and 'after' transfer periods of 1992 and 1996. The proportion of persons employed by the National Water Commission dropped from 100 per cent to 38 per cent. The associations feel that the Commission staff numbers are excessive and that: ...an unspecified percentage of these...staff are residual personnel that for political and labour-union-related reasons remain within the agency with no specific task. This has been one of the major reasons why the modules wanted to create the [District Federation of the user associations] which will take over management of the main system. They feel that this will be much more cost-effective than the current arrangement.

As already indicated, the water user associations hire all their own staff. In doing so they have shown reluctance to employ ex-Commission personnel. The associations argue that Commission channel-keepers and other staff were often out of control, poor workers, unaccountable, given to rent-seeking behaviour and were likely to involve the trade unions in module management in a divisive manner.

To what degree do farmers accept that the shared use of the resource is now equitable? The importance of the new, district-wide, hydraulic committees should be noted, but the allocation planning method has not changed at this level. However, there is evidence that correspondence between the volume of water assigned to modules and the actual volumes received by them is better than before transfer. This is probably because every association was represented on the hydraulic committee, with each module insisting on receiving the volume it had been assigned and had paid for. With respect to individual farmers, a detailed, post-transfer study shows that in the sample area 'there is no clear bias towards head- or tail-end farmers and that all farmers receive sufficient water to meet crop requirements'.

In a sample of farmers, 34 per cent said water distribution amongst them was poor before transfer and good after it, compared with only 15 per cent who held the reverse opinion. Probably the most important contribution to equitable distribution of the resource has come from the associations' control over channel-keepers. Bribes exacted by Commission employees under the previous regime were impossible for farmers to eradicate. The downward shift in power to water users now gives them the control over rent-seeking behaviour that they lacked. *To what degree has infrastructural operation and maintenance been improved?* With respect to operation, the post-transfer study indicated an improvement in water adequacy at field level, in the timeliness of water delivery and in access to the channel-keepers. In the sample of farmers, 40 per cent said the service provided by the channel-keepers was poor before transfer and good after it, compared with only 14 per cent who held the reverse opinion. Eighty-three per cent of farmers believe that the water user associations should now retain responsibility for operation of the main and secondary systems.

The transfer process included free concessionary use of Commission machinery such as draglines and hydraulic excavators. This was a flying start for users. Significantly, some of this equipment was in serious disrepair so the modules bought twenty-nine pieces of new heavy machinery from their user fee income and also received equipment from a World Bank programme.

Maintenance expenditure in constant peso prices almost doubled in the before- and after-transfer comparison. There was an approximately threefold increase in the desilting of secondary canals and drains. Farmers' perceptions of the condition of the irrigation and drainage network were that the network had improved considerably.

To what degree have negative externalities arising from individual farmars' actions declined since the user associations were set up? Kloezen and his colleagues address this question only with respecp to overabstraction of groundwater. By customary practice, modules exclude from the surface-water supply those areas that have access to wells. But groundwater users are permitted to use the canal network to distribute their pumped supplies.

Unfortunately the aquifers in the district are definitely being overexploited by about 20 per cent of annual recharge. By 1996 the water table was falling by 2-5 m annually. Groundwater mining does not appear to have become worse since transfer, but transfer has been unable to diminish it. To conclude this case study, the new arrangements for the financing of capital and current costs are examined. We have already seen that the association pays fees to the Commission for its services, that the association charges the farmer for each irrigation, that there has been a decline in bribes paid to channel-keepers and that charges to farmers have been used not only for prime costs but also for equipment purchase. Forty per cent of farmers surveyed report that payment procedures are now less cumbersome compared with only 2 per cent holding the reverse opinion.

Sixty-nine per cent stated that bribery of channel-keepers had been reduced. But many farmers still come to the module office to complain of keepers demanding private payments and this has often led to workers being dismissed. Keepers are also rotated between module sections to prevent the growth of patron-client relations. Nevertheless as many as 30 per cent of farmers surveyed were willing to admit that they bribe keepers to irrigate more than the farmer's entitled area or to get water at different times from those programmed. So the indirect charges are higher than the official ones- as are the wages of keepers. Under the Commission, charges were way below prime costs, let alone the full cost of the service. To prepare the way for a more self-reliant organization, the Commission increased the charge by some 400 per cent 2 years prior to transfer. It then remained unchanged in nominal pesos through and after transfer. But inflation halved the real value of the charge in the first 2 years of the associations' existence.

What is undoubtedly impressive is that in terms of the ratio of fees actually collected to actual OM&M expenditure, both given in pesos at current prices, the rate jumped from an average of 50 per cent in 1989-91 to 123 per cent in 1993-6. These facts remind us that in irrigation economics the estimation of the cost of water to farming families and agribusinesses should be based on the payments *actually* made, not on those that are *supposed* to be made. Kloezen and his colleagues never attempt to measure what proportion of the full cost of irrigation services is met by farmers' actual payments, but it seems likely to be considerably less than 100 per cent. This raises the question: 'What organisation in future will be responsible for infrastructural investment and rehabilitation, how will it raise the required finance and how will these costs be repaid?'

For the 1996-7 winter season the associations used the hydraulic committee to raise the irrigation fee by more than one-third. They also pushed the Commission to transfer OM&M responsibility for the main canal to a newly created District Federation of the user associations. In 1997 the average percentage of farmers' irrigation fees paid to the Commission was correspondingly slashed from 25 per cent to 9.5 per cent.

SELECTION OF IRRIGATION METHODS

Under normal conditions, the type of irrigation method selected will depend on water supply conditions, climate, soil, crops to be grown, cost of irrigation method and the ability of the farmer to manage the system. However, when using wastewater as the source of irrigation other factors, such as contamination of plants and harvested product, farm workers, and the environment, and salinity and toxicity hazards, will need to be considered. There is considerable scope for reducing the undesirable effects of wastewater use in irrigation through selection of appropriate irrigation methods.

The choice of irrigation method in using wastewater is governed by the following technical factors:

- The choice of crops,
- The wetting of foliage, fruits and aerial parts,
- The distribution of water, salts and contaminants in the soil,
- The ease with which high soil water potential could be maintained,
- The efficiency of application, and
- The potential to contaminate farm workers and the environment.

A border (and basin or any flood irrigation) system involves complete coverage of the soil surface with treated effluent and is normally not an efficient method of irrigation. This system will also contaminate vegetable crops growing near the ground and root crops and will expose farm workers to the effluent more than any other method. Thus, from both the health and water conservation points of view, border irrigation with wastewater is not satisfactory.

The efficiency of surface irrigation methods in general, borders, basins, and furrows, is not greatly affected by water quality, although the health risk inherent in these systems is most certainly of concern. Some problems might arise if the effluent contains large quantities of suspended solids and these settle out and restrict flow in transporting channels, gates, pipes and appurtenances. The use of primary treated sewage will overcome many of such problems. To avoid surface ponding of stagnant effluent, land levelling should be carried out carefully and appropriate land gradients should be provided.

Sprinkler, or spray, irrigation methods are generally more efficient in terms of water use since greater uniformity of application can be achieved. However, these overhead irrigation methods may contaminate ground crops, fruit trees and farm workers. In addition, pathogens contained in aerosolized effluent may be transported downwind and create a health risk to nearby residents. Generally, mechanized or automated systems have relatively high capital costs and low labour costs compared with manually-moved sprinkler systems. Rough land levelling is necessary for sprinkler systems, to prevent excessive head losses and achieve uniformity of wetting.

Sprinkler systems are more affected by water quality than surface irrigation systems, primarily as a result of the clogging of orifices in sprinkler heads, potential leaf burns and phytotoxicity when water is saline and contains excessive toxic elements, and sediment accumulation in pipes, valves and distribution systems. Secondary wastewater treatment has generally been found to produce an effluent suitable for distribution through sprinklers, provided that the effluent is not too saline. Further precautionary measures, such as treatment with granular filters or micro-strainers and enlargement of nozzle orifice diameters to not less than 5 mm, are often adopted.

Localized irrigation, particularly when the soil surface is covered with plastic sheeting or other mulch, uses effluent more efficiently, can often produce higher crop yields and certainly provides the greatest degree of health protection for farm workers and consumers. Trickle and drip irrigation systems are expensive, however, and require a high quality of effluent to prevent clogging of the emitters through which water is slowly released into the soil.

When compared with other systems, the main advantages of trickle irrigation seem to be:

- Increased crop growth and yield achieved by optimizing the water, nutrients and air regimes in the root zone,

- High irrigation efficiency - no canopy interception, wind drift or conveyance losses and minimal drainage losses,
- Minimal contact between farm workers and effluent,
- Low energy requirements - the trickle system requires a water pressure of only 100-300 k Pa (1-3 bar),
- Low labour requirements - the trickle system can easily be automated, even to allow combined irrigation and fertilization (sometimes terms fertigation).

Apart from the high capital costs of trickle irrigation systems, another limiting factor in their use is that they are only suited to the irrigation of row crops. Relocation of subsurface systems can be prohibitively expensive. Clearly, the decision on irrigation system selection will be mainly a financial one but it is to be hoped that the health risks associated with the different methods will be taken into account.

The method of effluent application is one of the health control measures possible, along with crop selection, wastewater treatment and human exposure control. Each measure will interact with the others and thus a decision on irrigation system selection will have an influence on wastewater treatment requirements, human exposure control and crop selection (for example, row crops are dictated by trickle irrigation). At the same time the irrigation techniques feasible will depend on crop selection and the choice of irrigation system might be limited if wastewater treatment has already been decided before effluent use is considered.

PARTICIPATORY IRRIGATION MANAGEMENT

The underlying theme of this chapter is that accumulated experience shows the need for a thorough rethinking of approaches to participatory irrigation management, synthesizing the best aspects of past programmes into a more efficient and effective approach to PIM. Many valuable lessons have emerged from the implementation of participatory programmes. The sustainability of participation, as currently developed, is highly questionable. Better implementation, improved coordination, perfecting current methods, stronger high level support and increased efforts may be useful, but by themselves are unlikely to bring about the changes needed to make participation sustainable and worthwhile for farmers. The challenge lies in reengineering participation in irrigation operation and maintenance processes to be much more efficient, better aligned with farmers' incentives and more effective in achieving the goals of improving irrigated agriculture and the welfare of farmers.

Turnover of Small Irrigation Systems to WUA

Participatory planning of physical improvements and registration of formal WUA are used to prepare for turnover of irrigation systems smaller than 500 hectares to water user associations. Irrigation inspectors are trained to facilitate local participation in the turnover process. Overall procedures for turnover were laid out in an ordinance from the Minister of Public Works (No. 42/PRT/1989).

The key steps of inventory/socio-technical profiling, design, construction of improvements, WUA derelopment and turnover. The government decides which systems will be turned over, so inclusion in the programme is based on government programming rather than responding to local requests. Implementation is now primarily handled by functional (project) units of the Provincial Water Resources Services (PWRS), under the guidance of the Directorate General of Water Resources Development (DGWRD), Ministry of Public Works.

Pilot activities began in 1987 and programme implementation in 1988. The first systems were officially turned over in 1990. By 1996, systems irrigating over 200,000 hectares had been turned over, or were fully prepared, awaiting only completion of administrative formalities. The remaining area served by irrigation systems smaller than 500 hectares is estimated to be over 700,000 hectares.Farmer incentives: Farmers gain government assistance for physical improvements. Assistance for building and repairing weirs, division structures, canal lining and other improvements has averaged from one to two hundred dollars per hectare. Farmers gain government recognition of their role in management and clearer authority for management. WUA registration formally legitimizes local management institutions and may give greater authority in enforcing rules for water distribution and mobilization of resources.

Farmers lose government assistance in routine operation and maintenance. However many systems did not receive any such assistance. In many other systems such assistance constituted only minor funding for cutting grass and desilting, or occasional funding for minor repairs. In many small systems which were turned over, farmers were not even aware that the government considered their system to be a "government" system, perhaps based on a few structures built with government funding many years ago.

Impacts of Turnover on Production and Farmer Welfare

There is a shortage of information on the impacts of turnover on production and farmer welfare. A 1994 study of 50 randomly selected turnover and starter project sites in West Java and East Nusa Tenggara indicated that on average yields increased one quarter ton per hectare per season, from more reliable and more equitably distributed water supplies. Overall economic rates of return were high, but about one quarter of sites had poor returns, showing that positive impacts are not certain, but require good site selection and planning.

Government Goals

Turnover of O&M responsibilities was intended to reduce demands on government resources, which could then be reallocated, particularly to help provide more adequate staff and budget for larger government irrigation systems. More generally, turnover was intended to create a better partnership and division of responsibility between government and farmers. It would let the government concentrate its management efforts in larger systems, which serve the majority of irrigated areas, while WUA would fully manage smaller systems, which are much larger in number but cover only a minority of the irrigated area. Turnover, ISF and other O&M reforms were part of larger projects which allowed continuing access to international funding for irrigation system construction and rehabilitation.

Staffing Reduction

Staff reduction and reallocation may be benefits from a broad government perspective, but look negative from the point of view of individual bureaucratic units. Government budget allocations for O&M, particularly central government funds, are based on the irrigated area served by government (jointly-managed) irrigation systems, so, other things being equal, turnover would reduce the budget, a loss from the point of view of irrigation officials. Similarly, staff reductions and transfers potentially could shrink section and subsection irrigation offices. In practice, some weir caretakers and canal guards have been moved to other systems, released, or hired by WUA, though little systematic information seems available on staff reallocation. Irrigation inspectors have usually been left in place to serve other jointly-managed systems in their jurisdiction, and in theory supposed to assist farmers in turned over systems.

Implementation Issues of Turnover

Carrying out turnover on a nationwide basis has been a major undertaking. It involved development of new procedures in pilot areas, institutionalization in policies, manuals and training curricula, and extensive training and institutional change to carry out the new programme. This was particularly challenging for those aspects requiring new skills and attitudes, beyond civil engineering and construction project management.

Many of the problems are those which affect any large, new programme, including shortages of suitable staff, the need to train and retrain personnel, delays in availability of budgets, lack of synchronization between interrelated activities, and difficulties in adapting a national programme to the great diversity of local conditions. Much has been done to overcome such problems, though much potential still exists for improving the quality of implementation.

Hardware and Software Improvements

Turnover goals for physical improvements have been defined in terms of putting irrigation systems into improved condition before turnover. Decisions that all systems must have a formal WUA to accept turnover, and all should be eligible for construction assistance, have meant that almost all systems have improvements built before turnover. The pace of turnover has therefore been shaped by the cost and availability of funds for construction of physical improvements. There are grounds for concern that implementation of turnover is emphasizing physical construction works, without adequate emphasis on the institutional changes during and after turnover essential to sustain and improve performance in irrigation management.

Support After Turnover

According to policy, turnover is not intended to mean that farmers are abandoned, but that government aid should be for matters beyond the capacity of farmers. Formal WUA have organized in ways which assume there will at least be some minimal level of continuing support from government. Provincial budgets (APBD) for WUA guidance are very small and no funding is specifically itemized for guidance of WUA in turned over systems. There is not yet any programme for annual joint inspection of turned over irrigation systems, to provide technical advice, promote preventive maintenance and ensure that potentially expensive and dangerous problems beyond farmer capacity are identified and dealt with in a timely way. Some weirs which have failed during floods have been repaired with natural disaster funds from general budgets.

There is no insurance programme or systematic approach to budgeting for such assistance costs, nor for farmer-cost sharing in major repairs or improvements. Without post-turnover support, the goals of turnover will not be achieved and the project effort would largely be wasted. Adequate guidelines, resources and incentives are needed, if guidance and support are to be provided after turnover and the performance of turned-over systems sustained.

IRRIGATION SERVICE SUPPLY MANAGEMENT

The irrigation cycle's supply side consists of appropriating the five base water flows, their storage and distribution to and from storage through to farmers' fields. The economics of irrigation service supply-first, the making of the infrastructure and, second, the year-on-year delivery of water to farming families and agribusinesses. The Report of the World Commission on Dams (WCD) has made clear just how ancient are the infrastructural works of irrigation: The earliest evidence of river engineering is the ruins of irrigation canals over eight thousand years old in Mesopotamia. Remains of water storage dams found in Jordan, Egypt and other parts of the Middle East date back to at least 3000 BC. Historical records suggest that the use of dams for irrigation and water supply became more widespread about a thousand years later. At that time, dams were built in the Mediterranean region, China and Meso America. Remains of earth embankment dams built for diverting water to large community reservoirs can still be found in Sri Lanka and Israel. The Dujiang irrigation project, which supplied 800,000 hectares in China, is 2,200 years old.

The irrigation cycle draws on an immense variety of the economy's real resources both for infrastructural provision and for the subsequent annual operation and maintenance of the service. These resources include:

- Freshwater or brackish water at the point of appropriation;
- The land sites required for the system's infrastructure;
- Wells, boreholes, pumps and their power units such as combustion engines and electric motors;
- Reservoirs;
- Plant for mechanical, biological and chemical treatment of wastewater when treated wastewater is used for irrigation;
- An extensive system of canals, pipes and other channels;
- A wide variety of monitoring, measurement and control devices;
- Petrol and diesel fuel, electricity supplies and spare parts;
- The human resources necessary to design, build, operate, monitor, maintain, repair, rehabilitate, manage and administer the whole.

Headworks costs are those of abstraction and storage. Network costs are those of water distribution.

In the analysis of the supply of irrigation services, the economist converts the tabulation of these real resources into money expenditures, on the basis of the resources' market prices. Expenditure can then be divided into capital account spending and current account spending. *Capital expenditure* is defined as that where the resource purchased has an expected life of more than 12 months. This would include: land; civil engineering infrastructures such as boreholes, reservoirs, works road networks, pumps and pipes; buildings and plant of all kinds; and vehicles.

Current expenditure refers to purchased resources that are either immediately used up in the process of production or which have a life of 12 months or less. These include materials, chemicals, spare parts, electric power and labour time. In practice, relatively small expenditures, such as those on some types of office equipment, would be counted as current expenditure, even where their life exceeds 1 year. Where long-life resources are rented rather than purchased, or in other instances where payment is made on a regular basis rather than as a lump sum, the expenditure will be classified as a current account item. The rent of land and buildings, for example, will appear under current expenditure. From an economic and a financial perspective, the importance of the distinction between capital and current spending is that, *first,* investment in a newly developed system's civil engineering infrastructures and plant requires far greater expenditure on capital account than the current account costs of a single year's operations. As a result, capital spending often requires debt finance. *Second,* commitment to current account spending is more flexible than the sunk costs under capital account. *Third,* the balance between capital and current spending is of the greatest importance in project evaluation. *Fourth,* the profit and loss account of an agribusiness is made up only of current account incomes and expenditures. The profit and loss tabulation is the professional accountant's version of the simpler farm budget.

It has to be admitted that the current-capital distinction may be fuzzy. For example, the patching of a canal with low-quality concrete mortar to plug cracks doubtless would be classified as a current cost. But the replacement of a significant length of canal line with pukka concrete intended to last 10 years would fall under capital spending.

Real capital resource requirements for infrastructural supply[a] Appropriation of the five base flows Land sites; wells and boreholes; *qanat/ falaj* system headworks;[b] surface-water off-take structures; flood recession works;[c] positive displacement pumps; roto-dynamic pumps; power units; treatment works for waste water appropriated for supply; measurement and control devices; miscellaneous buildings, roads, vehicles and equipment; weed screens; gantry-mounted grabs; transformer units

StorageLand sites; aquifers for artificial recharge;[d] reservoirs, spillways, stilling basins, siphons; water tanks and cisterns; measurement and control devices; miscellaneous buildings, roads, vehicles and equipment Distribution Land sites; buried and surface pipes; lined and unlined canals; sluice-gates; measurement and control devices including cross-regulators, weirs, flumes, drop structures, stilling basins, siphons; hydraulic excavators, bulldozers, tractor-mounted mowers, draglines and skips; miscellaneous buildings, roads, vehicles and equipment

THE SHORT TERM: ECONOMIES OF CAPACITY UTILIZATION

Efficiency in the appropriation, storage, distribution and use of water for agriculture. In all cases but one these are *output efficiency* measures, having a crop output or crop sales measure as the numerator and a measure of irrigation water supply as the denominator. Of the first seven efficiency measures, only one is a supply-side ratio. This is *E*2 -net irrigation requirement divided by the base supply. *Supply-side efficiency* in terms of the supply cost per cubic metre of water delivered to the farmer's field. It is also important to remember that irrigation supply in cubic metres can be measured at many points. An irrigation engineer writes: In big systems there are very substantial losses before the water gets to the irrigation district boundary; there are more between that and the farm boundary and, where farms are big and open channel distribution systems in use, more losses before the water gets to the field. Finally, for the purpose of comparing alternatives for on-farm delivery, one should measure the supply both at the field boundary and at the plant. The two measures most often used in this text are *Sb*, base supply at the start of the irrigation cycle and *Sf*, the volume delivered to the farmer's fields. The most important concepts in the irrigation economist's tool-kit to explore supply-side efficiency are *cost functions*.

These quantitative relationships describe the cost of supplying output in any time-period at each scale of output from zero units up to the system's theoretical capacity. The ability to shift from one delivery level of irrigation water up to a higher level clearly depends on the time available to make the change. It was Alfred Marshall, one of the greatest British economists, who introduced periodization into the analysis of supply and demand. He distinguished between the short term and the long term. In its application to the irrigation cycle, this simple, twofold distinction can be represented in the following way: in the short term, increased daily output is possible through operational changes or by organizational innovations demanding new procedures, both of which are relatively straightforward in their introduction; in the long term, additional infrastructure is required either in new projects or for the expansion of existing capacity, with associated financial, resource and organizational planning.

In conventional analysis it is possible, for either the short or long term, to calculate the *total cost* of supply at each level of output per day. *Average cost* at each level of output can be derived from this by dividing the total cost by the number of units produced. The *marginal cost* of the *n*th unit of output is defined as the difference in total cost between producing n and $(n - 1)$ units. A related but alternative exposition is set out below. It has the advantage that the terms refer to engineering categories, making it easier for non-economists to grasp quickly. It is also more appropriate than the conventional analysis, since it is a better representation of the way irrigation institutions actually think about their costs. Expenditures are defined using three categories-the parallel with farming family and agribusiness costs will be evident:

- Prime costs are defined as those used up in the daily supply of irrigation services and consist of the salaries and wages of the workforce, the costs of power, materials, spare parts and other consumables such as bought-in specialist inputs.
- Overhead costs of production are non-prime costs and are often assumed to be invariant with the level of supply. In the simplest case, where capital expenditure on land, infrastructure and plant has been funded 100 per cent by loans, this element of annual overhead costs would be set equal to the annual interest payable and the principal repayable on the debt incurred. Where capital investment is funded, at least in part, from an irrigation authority's surplus of water charges

over prime costs, overhead costs include the amortization charges necessary in the long run to contribute to the rehabilitation and replacement of elements of the system infrastructure. The rent of land and buildings and the leasing costs of capital equipment such as vehicles, are also included in overheads.

- Total costs per year are equal to prime plus overhead costs.

For each level of output, one can calculate overhead cost per unit of output and this function, called average overhead cost, is represented for a hypothetical case, *illustrating the short-term situation*. Average overhead costs fall (at a decelerating rate) as the unchanged level of cost is divided by evergreater levels of production. So in the short term, supply-side efficiency in terms of the overhead costs of irrigation water supply, *i.e.* the supply costs of the headworks and networks infrastructure per unit of water delivered, is achieved by operating the irrigation system at its full capacity. In other words, at scales of delivery less than full capacity, the average cost of the appropriation, storage and distribution infrastructure necessary to supply irrigation water rises exponentially. Where an irrigation system is built to meet the peak period of demand-which may be only 10 days long-the system is operating at below capacity for most of the irrigation season. Alternatives are to design a system that does *not* meet peak demand or to use volumetric pricing to suppress the peak demand for water. In this respect Carruthers and Clark showed an interesting link between surface-water and groundwater abstraction;

Tubewells may be used where surface canal water is at present supplied to augment basic surface supplies and increase cropping intensity. Augmentation to meet relatively short periods of peak demand may be particularly valuable. The most profitable cropping pattern is unlikely to have an even profile of water demand over the year and a canal system is limited in its capacity to cope with peaks.

THE LONG TERM: ECONOMIES OF SCALE

We now switch to the long term. Again our interest is in how average overhead costs vary with output scale. In this case output is increased not by raising the volume of irrigation water supplied towards 100 per cent utilization of *a fixed* infrastructural capacity, but by *adding* to the infrastructural capacity through specified engineering works. We can say economies of scale exist where higher levels of output are associated with lower overhead cost per unit of irrigation water supplied.

For the long term, where output rises because of capacity increases resulting from the investment process, the best approach in understanding scale economies is through *ex ante* calculations, *i.e.* a view of future possibilities not yet realized-*a planning approach*. The starting point is a real catchment in a specific year, with all the infrastructural investment in irrigation services inherited from the past in place. The interest here is in average overhead cost at each alternative value of a set of supply volumes derived from additions to capacity, from a minimal increment through to any upper limit that one believes to be appropriate. Note that the curve does not consist of a set of points representing how average overhead cost changes as the scale of production expands (or contracts) over time. It is not an evolutionary function. What it represents, for each output level, is the average overhead cost if, in a single leap, capacity were expanded up to that level. It is referred to as a long-term supply function not because the figure represents economies of scale with the passage of time but because the long term is the conventional time-scale over which additions to capacity can take place, in contrast to the operational changes of the short term.

This is worth restating. Average overhead cost variations with output in the short run represent the cost-output relationship as management in real-time shifts output upwards or downwards from one level to another. Average overhead cost variations with output in the long run represent the cost-output relationship for each of a series of separate, mutually exclusive choices. Here it is logically impossible for there to be real-time shifts of output upwards or downward from one level to another. So, in the short term, shifts over time in output levels are represented; but in the long term, mutually exclusive choices between future additions to capacity and to output are represented. The hypothetical function clear economies of scale up to 140 units of output per day. However, here, diseconomies of scale appear above 150 units per day. This cost function is quadratic in nature, *i.e.* well represented by an equation ob the form:

$$\text{AOC} = L^2 + bL + c$$

where AOC is average overhead cost and L is the output quantity the equation is:

$$\text{AOC} = 0.004L^2 - 0.998L + 93.5$$

How can one explain long-term economies of scale in irrigation service provision? It is the measure of what economists widely refer to as the *indivisibility* of infrastructural provision. This neatly expresses the idea that, to secure even modest levels of output, major works are necessary-a canal is a good example-and these are negligibly less costly than works that secure more substantial output levels. As a result, AOC falls markedly. The existence of significant scale economies at low- tk medium-output levels has a most important social effect.

It implies that irrigation service supply constitutes what is widely called a natural monopoly. The same can hold true in infrastructural provision for fresh and wastewater services in general, as well as for transport, energy supply and telecommunications. The proposition that long-term economies of scale exist in the supply of irrigation services should be used more as a working hypothesis than as an accepted fact. Moreover, one can expect variation in the existence of such economies between the infrastructural categories of appropriation, storage and distribution. With respect to appropriation headworks, the existence or not of economies surely varies between the five base flows of the irrigation cycle. *Rainfall collection* using a harvesting surface and a cistern for storage can work well at the small scale and is used in many parts of the world by farming families for small plot cultivation, particularly for their subsistence needs. We saw such an example in the case study of the small agribusiness in Cornwall, where rainfall collection contributed one-sixth of the farm's irrigation requirements.

Irrigation water supplied by *the reuse of domestic and industrial wastewater* certainly can be expected to exhibit long-term scale economies in respect of any wastewater treatment prior to reuse. However, if the wastewater treatment plant's (WWTP) cost is accounted for as part of the domestic and industrial supply system, the average overhead cost of reuse water appropriation is zero. The reuse in farming of *agricultural drainage water* by means of pumps set in drainage channels seems to operate very effectively at small scales, so no long-term scale economies can be expected here. Surface-water and groundwater abstraction takes place on a wide variety of scales as we have already seen in the Tamil Nadu, Curu Valley and Cornwall case studies. I do not believe that one can expect to find a consistent pattern of long-term scale economies here. The variation in technology is phenomenal, from the treadle pump up to major infrastructures for river diversion. An extensive review of the use of the treadle pump in Africa shows how much more effective it is than traditional rope-and-bucket irrigation systems. In Bangladesh more than 500,000 of these low-cost devices are in daily use. Even where scale economies exist for one component of the irrigation cycle, such as a river barrage like that on the Bhavani River, distribution costs may be greater across the large area in command.

With the exception of springs and of surface-water diversion through civil engineering structures large or small, virtually all groundwater and surface-water abstraction is carried out by pumping. So the possible existence of long-term scale economies in pumps and their power units needs to be reviewed. Positive displacement pumps such as the piston pump, the rotary pump, the air-lift pump and the Archimedean screw are convenient for small discharges. Of the roto-dynamic pumps (centrifugal, axial and mixed flow) centrifugal pumps are the most widely used. They are ideally suited to small discharges at high pressures but in fact are very versatile.

Axial pumps are very efficient at lifting large volumes of water at low pressure; they tend to be used for large pumping works because of the expense of their drive shaft and bearings. For some middle-range head and discharge requirements, mixed flow pumps (combining the centrifugal and axial flow principles) perform best. Centrifugal pumps can also be operated in series when extra head is required. Pumps and their power units convert fuel energy such as diesel oil into useful water energy. One of the interests of hydraulic is how well the power from the engine or motor is converted into useful water power in the pump. This is given as a measure known as power efficiency. In the absence of friction, efficiency would be 100 per cent but there are always friction losses in all the components of both the pump and its engine. Economies of scale *do* exist here. As Melvyn Kay writes: 'Smaller pumps are usually less efficient than larger ones because there is more friction to overcome relative to their size.'

Force-mode in contrast with suction-mode pump technologies are more expensive but give access to groundwater at greater depth. Here, in particular, the larger investment leads purchasers to give greater consideration to pump and engine performance, reliability, repair facilities and spare parts availability. With respect to power units, electric motors are very efficient in energy use and can be employed to drive all sizes and types of pump. Where access to electricity is not easy or is unreliable, petrol or diesel engines are used. Diesel engines are heavier and more expensive to purchase but they are more robust than petrol engines, have a much longer working life and are cheaper to run.

Turning now to storage head works, strictly in terms of their engineering economics, dams certainly offer economies of scale. For example, large dams in terms of storage capacity on balance have a greater depth of water and therefore lose less of their water to evaporation. Again, as Melvyn Kay points out in his excellent introduction to hydraulics, the thickness and therefore the cost of a dam wall's construction is determined by the head of water it contains, not by the volume of water it stores. This is known as the dam paradox. In general, the ratio of the number of cubic metres of water stored to each cubic metre of wall increases with dam size. The empirical relation between dam size and capital cost for each thousand cubic metres stored strongly confirms the existence of scale economies.

The costs of storage are also strongly affected by site conditions: in broad valleys with poor foundation conditions, costs will be high. An ideal site is a narrow gorge broadening to a large valley up-stream with good, solid rock conditions and no risk of earthquake activity. In the case of storage, as with other infrastructural works, the best sites tend to be developed first. Max Neutze stresses that unambiguous scale economies exist for the *networks* of land drainage, flood control and fresh and wastewater services, including irrigation service supply.

The cost of pipes increases roughly in proportion to their length and diameter, but their capacity increases in proportion to their diameter to the power of about 2.6. This is because the cross-sectional area increases as the square of the diameter and because friction between the water and the pipe decreases with size. Indivisibility, Neutze argues, is a characteristic of networks. On the subject of networks, an advantage of boreholes over surface-water surfaces is that they can be sited adjacent to the areas to be irrigated, thus reducing distribution overheads. Stressed that the long-term cost function deals with additions to capacity 'in a single leap', that evolutionary development is not in question here. However, when engineers are preparing plans many years ahead, it may be necessary to plot an evolutionary path. In the case of a 10-year water strategy, for example, it may make sense to consider two rounds of investment.

Here one is concerned not with a once-for-all heave to arrive at a given output, as with the long-term supply function, but with a first investment commitment followed by a second round of construction, for example, as an agricultural area's demand expands over time. Average overhead cost for the first of the two investment rounds in our example. But the AOC of the second round is not so defined, since the long-term cost function will have shifted as a result of the prior introduction of the first-round infrastructures.

Here is an example. Suppose that the stock of water in a reservoir for an irrigation district when it is fully developed will need to be 100 units. The single-round solution might simply be to build a 100-unit facility. A two-round solution might be to construct a 75-unit reservoir initially and then to add 25 units more in a second stage. The single-throw approach may have the advantage of providing a more cost-effective solution when the district is fully in production but at the cost of underutilized capacity during the first phase of the district's development.

But note, too, that the second-round addition of 25 units might be achieved either by a free-standing 25-unit reservoir or by deepening the 75-unit facility. With a double-throw, the first investment round changes the options available when the second round arrives. So, we can define the *indivisibility* of water infrastructures as what makes a single 20-m dam cheaper than two dams each of 10 m-a cross-section comparison. The *lumpiness* of water infrastructures is what makes a 10-m dam built now impose a higher construction cost for an additional 10-m capacity a year later-a time-series comparison. Lumpiness particularly characterizes the irrigation networks of pipes and canals. Lumpiness occurs because land prices will have been pushed up by first-round infrastructural construction, because pipe and canal installation is more expensive when it necessitates digging up an existing distributional infrastructure and because second-round investment is more disruptive than that on virgin sites.

If the range of second-round options are part of a long-term infrastructural plan in designing the first-round facility, the 25-unit expansion is likely to be achieved more cost-effectively than without such planning. This is the basis of Neutze's argument that the durability, specialization, immobility, indivisibility and lumpiness of infrastructural provision provide a strong economic argument for land-use planning and the demand management of growth. Note that groundwater abstraction is far less lumpy than surface-water abstraction. Tubewell construction can be phased in with demand, unlike canal systems and dams. The attractions of short-term economies of capacity utilization contrasted with the potential advantages of long-term planning and the link between prime and overhead costs, are illustrated by Kay: Pipe sizes are often selected using a discharge based on present water demands and little thought is given to how this might change in the future. Also there is always a temptation to select small pipes to satisfy current demand simply because they are less expensive than larger ones. These two factors can lead to trouble in the future. If demand increases and higher discharges are required from the same pipe, the energy losses will rise sharply and so a lot more energy is needed to run the system.

As an example, a 200 mm diameter pumped pipeline, 500 m long supplies a (discharge) of 50 litre/s. Several years later the demand doubles to 100 litres/s. This increases the velocity in the pipe from 1.67 to 3.34 m/s (*i.e.* it doubles) and the energy loss rises from 5 m to 20 m (*i.e.* a four-fold increase). This increase in head loss plus the extra flow means that eight times more energy is needed to operate the system and extra pumps may be needed to provide the extra power required. A little extra thought at the planning stage and a little more investment at the beginning could save a lot of extra pumping costs later. Note that Kay does not envisage adding a second 200-mm pipeline. The correct choice trading off short-term economies of utilization against lumpiness and long-term scale economies needs project evaluation over the life of the scheme. Infrastructural efficiency is not restricted to questions of scale economies. Capital costs per metre vary enormously. At the two extremes, buried asbestos pipes cost twenty-three times as much as high-density polyethylene (HDPE) surface piping-the distributional choice at Keveral Farm. There is little or no correlation between capital cost and either maintenance cost or distributional efficiency.

But the capital and maintenance costs of conveyance systems between the point of supply appropriation and the irrigated fields are not the only determinants of farmers' choices. Other factors of great importance are the *availability* of a distribution technology at the time of infrastructural construction, the *credit* provided for alternative technologies, the *permeability* of local soils and the hydraulic *head* available for water distribution, van Bentum and Smout, for example, point out that soils that are sandy or loamy favour buried pipe systems. Scarce supply combined with low hydraulic head favours lined channels. Compacted earth channels score best when soils are neither sandy nor loamy, where water is not scarce and where low head disfavours buried or partially buried pipe systems.

In hilly areas the cost of channels is increased because their alignment must follow the land's contours to create a gentle downward slope for the flow. A more direct but steeper route would bring higher flow velocity causing erosion and serious damage to the channels. Pipes can be used in any kind of terrain, can take a direct route and the resulting high water velocity brings no risk of soil erosion. Channels are particularly attractive options for conveying large quantities of water over relatively flat land such as in large irrigation systems on river floodplains.

LAND AND SOIL MANAGEMENT

Several land and soil management practices can be adopted at the field level to overcome salinity, sodicity, toxicity and health hazards that might be associated with the use of treated wastewater.

Land development

During the early stages of on-farm land development, steps can be taken to minimize potential hazards that may result from the use of wastewater. These will have to be well planned, designed and executed since they are expensive and, often, one time operations. Their goal is to improve permanently existing land and soil conditions in order to make irrigation with wastewater easier. Typical activities include levelling of land to a given grade, establishing adequate drainage (both open and sub-surface systems), deep ploughing and leaching to reduce soil salinity.

Land grading

Land grading is important to achieve good uniformity of application from surface irrigation methods and acceptable irrigation efficiencies in general. If the wastewater is saline, it is very important that the irrigated land is appropriately graded. Salts accumulate in the high spots which have too little water infiltration and leaching, while in the low spots water accumulates, causing waterlogging and soil crusting. Land grading is well accepted as an important farm practice in irrigated agriculture. Several methods are available to grade land to a desired slope. The slope required will vary with the irrigation system, length of run of water flow, soil type, and the design of the field. Recently, laser techniques have been applied to level land precisely so as to obtain high irrigation efficiencies and prevent salinization.

Deep cultivation

In certain areas, the soil is stratified, and such soils are difficult to irrigate. Layers of clay, sand or hard pan in stratified soils frequently impede or prevent free movement of water through and beyond the root zone. This will not only lead to saturation of the root zone but also to accumulation of salts in the root zone. Irrigation efficiency as well as water movement in the soil can be greatly enhanced by sub-soiling and chiselling of the land. The effects of sub-soiling and chiselling remain for about 1 to 5 years but, if long term effects are required, the land should be deep and slip ploughed. Deep or slip ploughing is costly and usually requires the growing of annual crops soon after to allow the settling of the land. Following a couple of grain crops, grading will be required to re-establish a proper grade to the land.

CROP MANAGEMENT AND CULTURAL PRACTICES

Several cultural and crop management practices that are valid under saline water use will be valid under wastewater use. These practices are aimed at preventing damage to crops caused by salt accumulation surrounding the plants and in the root zone and adjusting fertilizer and agrochemical applications to suit the quality of the wastewater and the crop.

Placement of seed

In most crops, seed germination is more seriously affected by soil salinity than other stages of development of a crop. The effects are pronounced in furrow-irrigated crops, where the water is fairly to highly saline. This is because water moves upwards by capillarity in the ridges, carrying salts with it. When water is either absorbed by roots or evaporated, salts are deposited in the ridges. Typically, the highest salt concentration occurs in the centre of the ridge, whereas the lowest concentration of salt is found along the shoulders of the ridges. An efficient means of overcoming this problem is to ensure that the soil around the germinating seeds is sufficiently low in salinity. Appropriate planting methods, ridge shapes and irrigation management can significantly decrease damage to germinating seeds. Some specific practices include:

- Planting on the shoulder of the ridge in the case of single row planting or on both shoulders in double row planting,
- Using sloping beds with seeds planted on the sloping side, but above the water line,
- Irrigating alternate rows so that the salts can be moved beyond the single seed row.

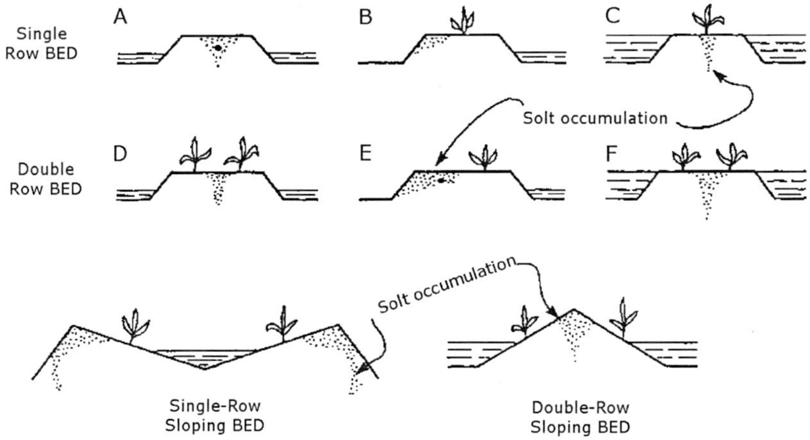

Fig. Schematic representations of salt accumulation and planting methods in ridge and furrow irrigation

WATER MANAGEMENT

Most treated wastewaters are not very saline, salinity levels usually ranging between 500 and 200 mg/l (EC_w = 0.7 to 3.0 dS/m). However, there may be instances where the salinity concentration exceeds the 2000 mg/l level. In any case, appropriate water management practices will have to be followed to prevent salinization, irrespective of whether the salt content in the wastewater is high or low.

It is interesting to note that even the application of a non-saline wastewater, such as one containing 200 to 500 mg/l, when applied at a rate of 20,000 m^3 per hectare, a fairly typical irrigation rate, will add between 2 and 5 tonnes of salt annually to the soil. If this is not flushed out of the root zone by leaching and removed from the soil by effective drainage, salinity problems can build up rapidly. Leaching and drainage are thus two important water management practices to avoid salinization of soils.

Leaching

The concept of leaching has already been discussed. The question that arises is how much water should be used for leaching, *i.e.* what is the leaching requirement? To estimate the leaching requirement, both the salinity of the irrigation water (EC_w) and the crop tolerance to soil salinity (EC_e) must be known. The necessary leaching requirement (LR) can be estimated from Figure for general crop rotations reported by Ayers and Westcot. A more exact estimate of the leaching requirement for a particular crop can be obtained using the following equation:

$$LR = \frac{EC_w}{5(EC_e - EC_w)}$$

where:

LR = minimum leaching requirement needed to control salts within the tolerance (EC_e) of the crop with ordinary surface methods of irrigation

EC_w = salinity of the applied irrigation water in dS/m

EC_e = average soil salinity tolerated by the crop as measured on a soil saturation extract. It is recommended that the EC_e value that can be expected to result in at least a 90 per cent or greater yield be used in the calculation. Figure was developed using EC_e values for the 90 per cent yield potential. For water in the moderate to high salinity range (>1.5 dS/m), it might be better to use the EC_e value for maximum yield potential (100 per cent) since salinity control is critical in obtaining good yields.

Where water is scarce and expensive, leaching practices should be designed to maximize crop production per unit volume of water applied, to meet both the consumptive use and leaching requirements. Depending on the salinity status, leaching can be carried out at each irrigation, each alternative irrigation or less frequently, such as seasonally or at even longer intervals, as necessary to keep the salinity in the soil below the threshold above which yield might be affected to an unacceptable level.

With good quality irrigation water, the irrigation application level will almost always apply sufficient extra water to accomplish leaching. With high salinity irrigation water, meeting the leaching requirement is difficult and requires large amounts of water. Rainfall must be considered in estimating the leaching requirement and in choosing the leaching method.

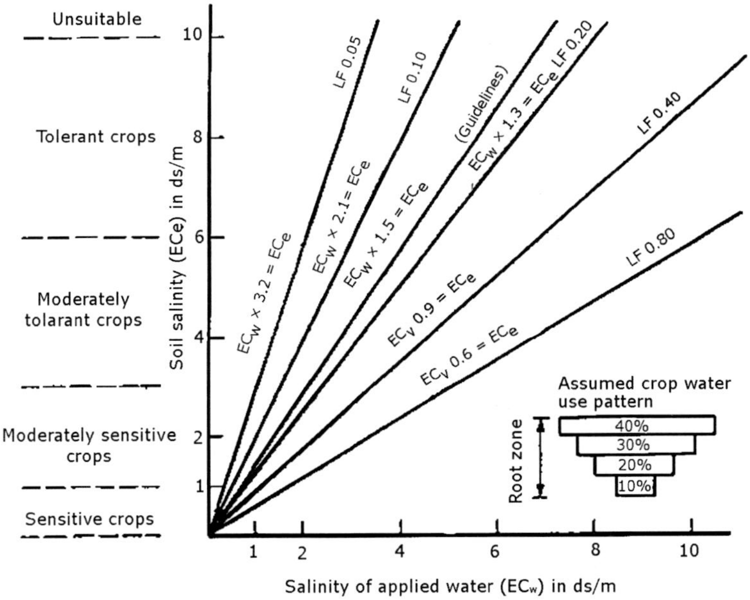

Fig. Relationship between applied water salinity and soil water salinity at different leaching fractions

The following practices are suggested for increasing the efficiency of leaching and reducing the amount of water needed:

- Leach during cool seasons instead of during warm periods, to increase the efficiency and ease of leaching, since the total annual crop water demand (ET, mm/year) losses are lower,
- Use more salt-tolerant crops which require a lower leaching requirement (LR) and thus have a lower water demand,
- Use tillage to slow overland water flow and reduce the number of surface cracks which bypass flow through large pores and decrease leaching efficiency,
- Use sprinkler irrigation at an application rate below the soil infiltration rate as this favours unsaturated flow, which is significantly more efficient for leaching than saturated flow. More

irrigation time but less water is required than for continuous ponding,

- Use alternate ponding and drying instead of continuous ponding as this is more efficient for leaching and uses less water, although the time required to leach is greater. This may have drawbacks in areas having a high water table, which allows secondary salinization between pondings,
- Where possible, schedule leachings at periods of low crop water use or postpone teachings until after the cropping season,
- Avoid fallow periods, particularly during hot summers, when rapid secondary soil salinization from high water tables can occur,
- If infiltration rates are low, consider pre-planting irrigations or off-season leaching to avoid excessive water applications during the crop season, and
- Use one irrigation before the start of the rainy season if total rainfall is normally expected to be insufficient for a complete leaching. Rainfall is often the most efficient leaching method because it provides high quality water at relatively low rates of application.

Drainage

Salinity problems in many irrigation projects in arid and semi-arid areas are associated with the presence of a shallow water table. The role of drainage in this context is to lower the water table to a desirable level, at which it does not contribute to the transport of salts to the root zone and the soil surface by capillarity.

What is important is to maintain a downward movement of water through soils. van Schilfgaard reported that drainage criteria are frequently expressed in terms of critical water table depths; although this is a useful concept, prevention of salinization depends on the establishment, averaged over a period of time, of a downward flux of water. Another important element of the total drainage system is its ability to transport the desired amount of drained water out of the irrigation scheme and dispose of it safely. Such disposal can pose a serious problem, particularly when the source of irrigation water is treated wastewater, depending on the composition of the drainage effluent.

Timing of irrigation

The timing of irrigation, including irrigation frequency, pre-planting irrigation and irrigation prior to a winter rainy season, can reduce the salinity hazard and avoid water stress between irrigations. Some of these practices are readily applicable to wastewater irrigation. In terms of meeting the water needs of crops, increasing the frequency of irrigation will be desirable as it eliminates water stress between irrigations. However, from the point of view of overall water management, this may not always produce the desired results. For example, with border, basin and other flood irrigation methods, frequent irrigations may result in an unacceptable increase in the quantity of water applied, decrease in water use efficiency and larger amounts of water to be drained. However, with sprinklers and localized irrigation methods, frequent applications with smaller amounts may not result in decrease in water use efficiency and, indeed, could help to overcome the salinity problem associated with saline irrigation water.

Pre-planting irrigation is practised in many irrigation schemes for two reasons, namely:

- to leach salts from the soil surface which may have accumulated during the previous cropping period and to provide a salt-free environment to germinating seeds (it should be noted that for most crops, the seed germination and seedling stages are most sensitive to salinity); and
- to provide adequate moisture to germinating seeds and young seedlings. A common practice among growers of lettuce, tomatoes and other vegetable crops is to pre-irrigate the field before planting, since irrigation soon after planting could create local water stagnation and wet spots that are not desirable. Treated wastewater is a good source for pre-irrigation as it is normally not saline and the health hazards are practically nil.

Blending of wastewater with other water supplies

One of the options that may be available to farmers is the blending of treated sewage with conventional sources of water, canal water or ground water, if multiple sources are available. It is possible that a farmer may have saline ground water and, if he has non-saline treated wastewater, could blend the two sources to obtain a blended water of acceptable salinity level. Further, by blending, the microbial quality of the resulting mixture could be superior to that of the unblended wastewater.

Alternating treated wastewater with other water sources

Another strategy is to use the treated wastewater alternately with the canal water or groundwater, instead of blending. From the point of view of salinity control, alternate applications of the two sources will be superior to blending. However, an alternating application strategy will require duel conveyance systems and availability of the effluent dictated by the alternate schedule of application.

CHALLENGES TO WATER MANAGEMENT INTEGRATION

The term "functional integration" means to join purposes of water management such as to manage water supply and wastewater within a single unit. Protecting aquatic habitat for natural and ecological systems while managing for flood control is another example. Still another term is "conjunctive use," which usually refers to the joint management of surface water and groundwater.

MANAGEMENT OF WATER QUALITY IMPACTS FROM FERTILIZERS

The investigation of eutrophication of surface water by agriculture must adopt a pragmatic management perspective. Of value to agriculture is the perspective adopted by the OECD study of eutrophication.

That study focused on the following aspects:

- The qualitative assessment of the trophic state of bodies of water in terms of a few easily measured parameters.
- The dependence of this state on nutritional conditions and nutrient load.
- Translation of these results to the needs of eutrophication control for management.

The progression of these aspects is interesting in that the focus is on easily measurable state of the water body, followed by a determination of the extent to which the state is a product of nutrient loads, then the degree to which loads may be manipulated to achieve a desired trophic state that is determined by water use.

Prediction of water quality impacts of fertilizers and related land management practices is an essential element of site-specific control options and for the development of generic approaches for fertilizer control. Prediction pools are essentially in the form of models.

Mineral Fertilizers

The response to the need to control leaching and run-off of nutrients and contamination of soils and water by heavy metals has been variable in Europe. Control measures are part of the larger issue of mineral and organic fertilizer usage.

FAO/ECE (1991) summarized the types of voluntary and mandated controls in Europe that apply to mineral fertilizers as:

- Taxes on fertilizer.
- Requirement for fertilizer plans.
- Preventing the leaching of nutrients after the growing season by increasing the area under autumn/winter green cover and by sowing crops with elevated nitrogen
- Promoting and subsidizing better application methods, developing new, environmentally sound fertilizers and promoting soil testing.
- Severely limiting the use of fertilizers in *e.g.* water extraction areas and nature protection areas.

In any location where intensive agriculture and/or livestock farming produces serious risks of nitrogen pollution, Ignazi (1993) recommended the following essential steps that are taken at the farm level:

Rational Nitrogen Application

To avoid over-fertilization, the rate of nitrogen fertilizer to be applied needs to be calculated on the basis of the "crop nitrogen balance". This takes into account plant needs and amount of N in the soil.

Vegetation Cover

As far as possible, keep the soil covered with vegetation. This inhibits build-up of soluble nitrogen by absorbing mineralized nitrogen and preventing leaching during periods of rain.

Manage the Period between Crops

Organic debris produced by harvesting is easily mineralized into leachable N. Steps to reduce leachable N includes planting of "green manure" crops and delaying ploughing of straw, roots and leaves into the soil.

Rational Irrigation

Poor irrigation has one of the worst impacts on water quality, whereas precision irrigation is one of the least polluting practices as well as reducing net cost of supplied water.

Optimize other Cultivation Rechniques

Highest yields with minimum water quality impacts require optimization of practices such as weed, pest and disease control, liming, balanced mineral fertilizers including trace elements, etc.

Agricultural Planning

Implement erosion control techniques that complement topographic and soil conditions.

Organic Fertilizers

Voluntary and legislated control measures in Europe are intended to have the following benefits:

- Reduce the leaching of nutrients
- Reducing emissions of ammonia
- Reducing contamination by heavy metals

The nature of these measures varies by country; however FAO/ECE have summarized the types of voluntary and mandated control as:

- Maximum numbers of animals per hectare based on amount of manure that can be safely applied per hectare of land.

- Maximum quantities of manure that can be applied on the land is fixed, based on the N and P content of the manure.
- Holdings wishing to keep more than a given number of animals must obtain a license.
- The periods during which it is allowed to apply manure to the land have been limited and it is obligatory to work it into the ground immediately afterwards.
- Establishment of regulations on minimum capacity for manure storage facilities.
- Establish fertilizer plans.
- Levies (taxes) on surplus manure.
- Areas under autumn/winter green cover were extended and green fallowing is being promoted.
- Maximum amounts established for spreading of sewage sludge on land based on heavy metal content.
- Change in composition of feed to reduce amount of nutrients and heavy metals.
- Research and implementation of means of reducing ammonia loss.

Sludge Management

Sludge is mentioned here only insofar as the spreading of sludge from municipal wastewater treatment facilities on agricultural land is one method used to get rid of municipal sludge in a way that is perceived to be beneficial. The alternatives are incineration and land fill. FAO/ECE include sludge within the category of organic fertilizers but note that sludge often contains unacceptable levels of heavy metals. Pollution of water by sludge run-off is otherwise the same as for manure.

ECONOMICS OF CONTROL OF FERTILIZER RUN-OFF

Nutrient loss is closely associated with rainfall-run-off events. For phosphorus, which tends to be associated with the solid phase (sediment), run-off losses are directly linked to erosion. Therefore the economics of nutrient control tend to be closely tied to the costs of controlling run-off and erosion. Therefore, this will be treated briefly here. In particular, it is useful to examine the economic cost of nutrient run-off which must be replaced by fertilizers if the land is to remain productive.

The link between erosion, increasing fertilizer application and loss of soil productivity is very direct in many countries. In the Brazilian state of Paraná where agriculture is the base of the state economy, Paraná produces 22 per cent of the national grain production on only 2.4 per cent of the Brazilian territory. Agricultural expansion in Paraná occurred mainly in the period 1950-1970 and was "characterized by short-term agricultural systems leading to continuous and progressive environmental degradation as a result of economic policies and a totally inappropriate land parcelling and marketing system...". Erosion has led to extensive loss of top soil, large-scale gullying and silting of ditches and rivers. The use of fertilizers has risen as a consequence, up 575 per cent over the period 1970-1986 and without any gain in crop yields. Loss of N-P-K from an average erosion of 20 t/ha/yr represents an annual economic loss of US$242 million in nutrients.

Analysis by Elwell and Stocking (1982) of nutrient loss arising from erosion in Zimbabwe shows similar significant economic losses in African situations. Stocking, applying data collected in the 1960s by Hudson to the soil use map of Zimbabwe, calculated an annual loss of 10 million tonnes of nitrogen and 5 million tonnes of phosphorus annually as a consequence of erosion.

Roose also cites losses of 98 kg/ha/yr of nitrogen, 29 kg/ha/yr of phosphorus, 39 kg/ha/yr of lime and 39 kg/ha/yr of magnesium from soils of lower Côte d'Ivoire as a result of erosion. This loss is so severe that compensation requires 7 tonnes of fresh manure annually, plus 470 kg of ammonium sulphate, 160 kg of superphosphate, 200 kg dolomite and 60 kg of potassium chloride per hectare per year.

Roose notes that it is not surprising that the soil is exhausted after only two years of traditional agriculture. Furthermore, these losses do not take into account additional losses due to harvesting and run-off. Roose summarizes by stating that action against soil erosion is essential in order to manage what he describes as a "terrible" chemical imbalance in soils caused by soil erosion. Economic losses tend to be higher in tropical countries where soils, rainfall and agricultural practices are more conducive to erosion and reported rates of erosion are much above average. The World Bank (1992) reported that extrapolation from test-plots of impacts of soil loss on agricultural productivity, indicates some 0.5-1.5 per cent loss of GDP annually for countries such as Costa Rica, Malawi, Mali and Mexico. These losses do not include offsite costs such as reservoir infilling, river sedimentation, damage to irrigation systems, etc.

Soil fertility is a complex issue and nutrient loss is not necessarily nor always a consequence of erosion. Erosion and soil loss is the end member of a variety of physical, vegetative and nutrient factors that lead to soil degradation. Global patterns of fertilizer application, as reported by Joly (1993), indicate however that rapidly rising levels of fertilizer utilization are required merely to maintain soil productivity from a variety of types of loss, including losses due to erosion and, more generally, to soil degradation. In a study of 17 agricultural sub-watersheds in the Lake Balaton district of Hungary, Jolankai (1986) measured and modelled N and P run-off from a variety of agricultural land uses. He calculated that a selection of control measures (mainly erosion control) would reduce phosphorus loss by 52.8 per cent at a cost of US$ 2500 per ha in remediation measures (in 1986).

AQUACULTURE

Aquaculture is a special case of agricultural pollution. There are two main forms: land-based and water-based systems. Effluent controls are possible on land-based systems, however water-based systems present particular problems. Aquaculture is rapidly expanding in most parts of the developed and developing world, both in freshwater and marine environments. In contrast, coastal fisheries in most countries are declining. The environmental impact is primarily a function of feed composition and feed conversion (faecal wastes), plus assorted chemicals used as biocides, disinfectants, medicines, etc. Wastage of feed (feed not taken up by the fish) is estimated to be 20 per cent in European aquaculture. Waste feed and faecal production both add substantial nutrient loadings to aquatic systems.

Additional environmental problems include risk of disease and disease transfer to wild fish, introduction of exotic species, impacts on benthic communities and on the eutrophication of water, interbreeding of escaped cultured fish with wild fish with consequent genetic change in the wild population. Traditional integrated aquaculture systems, as in China, where sewage-fish culture is practised, can be a stabilizing influence in the entire ecosystem. This is recommended, especially in developing countries where water and resources are scarce or expensive.

PROBLEMS OF RESTORATION OF EUTROPHIC LAKES

Eutrophic and hypertrophic lakes tend to be shallow and suffer from high rates of nutrient loadings from point and non-point sources. In areas of rich soils such as the Canadian prairies, lake bottom sediments are comprised of nutrient-enriched soil particles eroded from surrounding soils. The association of phosphorus with sediment is a serious problem in the restoration of shallow, enriched lakes. P-enriched particles settle to the bottom of the lake and form a large pool of nutrient in the bottom sediments that is readily available to rooted plants and which is released from bottom sediments under conditions of anoxia into the overlying water column and which is quickly utilized by algae.

This phosphorus pool, known as the "internal load" of phosphorus, can greatly offset any measures taken by river basin managers to control lake eutrophication by control of external phosphorus sources from agriculture and from point sources. Historically, dredging of bottom sediments was considered the only means of remediating nutrient-rich lake sediments, however, modern technology now provides alternative and more cost-effective methods of controlling internal loads of phosphorus by oxygenation and by chemically treating sediments *in situ* to immobilize the phosphorus. Nevertheless, lake restoration is expensive and must be part of a comprehensive river basin management programme.

PESTICIDES AS WATER POLLUTANTS

The term "pesticide" is a composite term that includes all chemicals that are used to kill or control pests. In agriculture, this includes herbicides (weeds), insecticides (insects), fungicides (fungi), nematocides (nematodes) and rodenticides (vertebrate poisons).

A fundamental contributor to the Green Revolution has been the development and application of pesticides for the control of a wide variety of insectivorous and herbaceous pests that would otherwise diminish the quantity and quality of food produce. The use of pesticides coincides with the "chemical age" which has transformed society since the 1950s. In areas where intensive monoculture is practised, pesticides were used as a standard method for pest control. Unfortunately, with the benefits of chemistry have also come disbenefits, some so serious that they now threaten the long-term survival of major ecosystems by disruption of predator-prey relationships and loss of biodiversity. Also, pesticides can have significant human health consequences.

While agricultural use of chemicals is restricted to a limited number of compounds, agriculture is one of the few activities where chemicals are intentionally released into the environment because they kill things. Agricultural use of pesticides is a subset of the larger spectrum of industrial chemicals used in modern society. The American Chemical Society database indicates that there were some 13 million chemicals identified in 1993 with some 500 000 new compounds being added annually. In the Great Lakes of North America, for example, the International Joint Commission has estimated that there are more than 200 chemicals of concern in water and sediments of the Great Lakes ecosystem. Because the environmental burden of toxic chemicals includes both agriculture and non-agricultural compounds, it is difficult to separate the ecological and human health effects of pesticides from those of industrial compounds that are intentionally or accidentally released into the environment. However, there is overwhelming evidence that agricultural use of pesticides has a major impact on water quality and leads to serious environmental consequences.

Irrigated agriculture, especially in tropical and subtropical environments, usually requires modification of the hydrological regime which, in turn, creates habitat that is conducive to breeding of insects such as mosquitoes which are responsible for a variety of vector-borne diseases. In addition to pesticides used in the normal course of irrigated agriculture, control of vector-borne diseases may require additional application of insecticides such as DDT which have serious and widespread ecological consequences. In order to address this problem, environmental management methods to control breeding of disease vectors are being developed and tested in many irrigation projects.

Historical Development of Pesticides

The history of pesticide development and use is the key to understanding how and why pesticides have been an environmental threat to aquatic systems and why this threat is diminishing in developed countries and remains a problem in many developing countries.

North-south Dilemma over Pesticide Economics

The general progression of pesticide development has moved from highly toxic, persistent and bioaccumulating pesticides such as DDT, to pesticides that degrade rapidly in the environment and are less toxic to non-target organisms. The developed countries have banned many of the older pesticides due to potential toxic effects to man and/or their impacts on ecosystems, in favour of more modern pesticide formulations. In the developing countries, some of the older pesticides remain the cheapest to produce and, for some purposes, remain highly effective as, for example, the use of DDT for malaria control.

Developing countries maintain that they cannot afford, for reasons of cost and/or efficacy, to ban certain older pesticides. The dilemma of cost/efficacy versus ecological impacts, including long range impacts via atmospheric transport and access to modern pesticide formulations at low cost remains a contentious global issue. In addition to ecological impacts in countries of application, pesticides that have been long banned in developed countries (such as DDT, toxaphene, etc.), are consistently found in remote areas such as the high arctic.

Chemicals that are applied in tropical and subtropical countries are transported over long distances by global circulation. The global situation has deteriorated to the point where many countries are calling for a global convention on "POPs" (Persistent Organic Pollutants) which are mainly chlorinated compounds that exhibit high levels of toxicity, are persistent and bioaccumulate. The list is not yet fixed; however, "candidate" substances include several pesticides that are used extensively in developing countries

FATE AND EFFECTS OF PESTICIDES

Human Health Effects of Pesticides

Perhaps the largest regional example of pesticide contamination and human health is that of the Aral Sea region. UNEP (1993) linked the effects of pesticides to "the level of oncological (cancer), pulmonary and haematological morbidity, as well as on inborn deformities... and immune system deficiencies".

Human health effects are caused by:

- *Skin contact:* Handling of pesticide products
- *Inhalation:* Breathing of dust or spray

- *Ingestion:* Pesticides consumed as a contaminant on/in food or in water.

Farm workers have special risks associated with inhalation and skin contact during preparation and application of pesticides to crops. However, for the majority of the population, a principal vector is through ingestion of food that is contaminated by pesticides. Degradation of water quality by pesticide run-off has two principal human health impacts. The first is the consumption of fish and shellfish that are contaminated by pesticides; this can be a particular problem for subsistence fish economies that lie downstream of major agricultural areas. The second is the direct consumption of pesticide-contaminated water. WHO (1993) has established drinking water guidelines for 33 pesticides. Many health and environmental protection agencies have established "acceptable daily intake" (ADI) values which indicate the maximum allowable daily ingestion over a person's lifetime without appreciable risk to the individual.

Ecological Effects of Pesticides

Pesticides are included in a broad range of organic micro pollutants that have ecological impacts. Different categories of pesticides have different types of effects on living organisms, therefore generalization is difficult. Although terrestrial impacts by pesticides do occur, the principal pathway that causes ecological impacts is that of water contaminated by pesticide run-off. The two principal mechanisms are bioconcentration and biomagnification.

Bioconcentration

This is the movement of a chemical from the surrounding medium into an organism. The primary "sink" for some pesticides is fatty tissue ("lipids"). Some pesticides, such as DDT, are "lipophilic", meaning that they are soluble in and accumulate in, fatty tissue such as edible fish tissue and human fatty tissue. Other pesticides such as glyphosate are metabolized and excreted.

Biomagnification

This term describes the increasing concentration of a chemical as food energy is transformed within the food chain. As smaller organisms are eaten by larger organisms, the concentration of pesticides and other chemicals are increasingly magnified in tissue and other organs. Very high concentrations can be observed in top predators, including man. The ecological effects of pesticides (and other organic contaminants) are varied and are often inter-related. Effects at the organism or ecological level are usually considered to be an early warning indicator of potential human health impacts.

The major types of effects are listed below and will vary depending on the organism under investigation and the type of pesticide. Different pesticides have markedly different effects on aquatic life which makes generalization very difficult. The important point is that many of these effects are chronic (not lethal), are often not noticed by casual observers, yet have consequences for the entire food chain.

- Death of the organism.
- Cancers, tumours and lesions on fish and animals.
- Reproductive inhibition or failure.
- Suppression of immune system.
- Disruption of endocrine (hormonal) system
- Cellular and DNA damage.
- Teratogenic effects (physical deformities such as hooked beaks on birds).
- Poor fish health marked by low red to white blood cell ratio, excessive slime on fish scales and gills, etc.
- Intergenerational effects (effects are not apparent until subsequent generations of the organism).
- Other physiological effects such as egg shell thinning.

These effects are not necessarily caused solely by exposure to pesticides or other organic contaminants, but may be associated with a combination of environmental stresses such as eutrophication and pathogens. These associated stresses need not be large to have a synergistic effect with organic micro pollutants.

Ecological effects of pesticides extend beyond individual organisms and can extend to ecosystems. Swedish work indicates that application of pesticides is thought to be one of the most significant factors affecting biodiversity. Jonsson report that the continued decline of the Swedish partridge population is linked to changes in land use and the use of chemical weed control. Chemical weed control has the effect of reducing habitat, decreasing the number of weed species and of shifting the balance of species in the plant community. Swedish studies also show the impact of pesticides on soil fertility, including inhibition of nitrification with concomitant reduced uptake of nitrogen by plants. These studies also suggest that pesticides adversely affect soil micro-organisms which are responsible for microbial degradation of plant matter (and of some pesticides) and for soil structure.

NATURAL FACTORS THAT DEGRADE PESTICIDES

In addition to chemical and photochemical reactions, there are two principal biological mechanisms that cause degradation of pesticides.

These are:

1. Microbiological processes in soils and water
2. Metabolism of pesticides that are ingested by organisms as part of their food supply.

While both processes are beneficial in the sense that pesticide toxicity is reduced, metabolic processes do cause adverse effects in, for example, fish.

Energy used to metabolize pesticides and other xenobiotics (foreign chemicals) is not available for other body functions and can seriously impair growth and reproduction of the organism.

Degradation of Pesticides in Soil

"Many pesticides dissipate rapidly in soils. This process is mineralization and results in the conversion of the pesticide into simpler compounds such H_2O, CO_2 and NH_3. While some of this process is a result of chemical reactions such as hydrolysis and photolysis, microbiological catabolism and metabolism is usually the major route of mineralization. Soil micro biota utilize the pesticide as a source of carbon or other nutrients. Some chemicals (for example 2,4-D) are quite rapidly broken down in soil while others are less easily attacked (2,4,5-T). Some chemicals are very persistent and are only slowly broken down (atrazine)".

Process of Metabolism

Metabolism of pesticides in animals is an important mechanism by which organisms protect themselves from the toxic effects of xenobiotics (foreign chemicals) in their food supply. In the organism, the chemical is transformed into a less toxic form and either excreted or stored in the organism. Different organs, especially the liver, may be involved, depending on the chemical. Enzymes play an important role in the metabolic process and the presence of certain enzymes, especially "mixed" function oxygenases (MFOs) in liver, is now used as an indicator that the organism has been exposed to foreign chemicals.

GOVERNMENTAL AND INTEREST GROUPS

Accommodating the views of governments and special interest groups is a challenge in integration because they have different perspectives. Intergovernmental relationships between government agencies at the same level include regional, state-to-state and interagency issues. Relationships between different levels of government include, for example, state–federal and local–state interactions. Special interest groups range from those favouring development of resources to those favouring preservation. In many cases, conflicts arise between the same types of interest groups, as, for example, between fly fishers and rafters on a stream.

Geographic Regions

The views of stakeholders in different locations must be balanced, introducing a geographic dimension of integration. Examples include issues between upstream and downstream stakeholders, issues among stakeholders in the same region and views of stakeholders in a basin of origin versus those in a receiving basin. Another aspect of geographic integration is the scale of water-accounting units, such as small watershed, major river basin, region, or state, even up to global scale.

Interdisciplinary Perspectives

The complexity of integrated water resources management requires knowledge and wisdom from different areas of knowledge, or disciplines. Blending knowledge from engineering, law, finance, economics, politics, history, sociology, psychology, life science, mathematics and other fields can bring valuable knowledge about the possibilities and consequences of decisions and actions. For example, engineering knowledge might focus on physical infrastructure systems, whereas sociology or psychology might focus on human impacts.

COORDINATION AND COOPERATION

Coordination is an important tool of integration because the arena of water management sometimes involves conflicting objectives. Coordinating mechanisms can be formal, such as intergovernmental agreements, or informal, such as local watershed groups meeting voluntarily.Cooperation is also a key element in integration, whether by formal or by informal means. Cooperation can be any form of working together to manage water, such as in cooperative water management actions on a regional scale, often known as "regionalization."

Examples of regionalization include a regional management authority, consolidation of systems, a central system acting as water wholesaler, joint financing of facilities, coordination of service areas, interconnections for emergencies and sharing of personnel, equipment, or services.

Total Water Management

Integrated water resources management can take different forms and is examined best in specific situations. In the water-supply field, the term "integrated resource planning" has come into use to express concepts of integration in supply development. Perhaps the most comprehensive concept for water supply is "Total Water Management."

According to a 1996 report of the American Water Works Research Foundation, Total Water Management is the exercise of stewardship of water resources for the greatest good of society and the environment. A basic principle of Total Water Management is that the supply is renewable, but limited and should be managed on a sustainable-use basis.

Taking into consideration local and regional variations, Total Water Management:

- Encourages planning and management on a natural water systems basis through a dynamic process that adapts to changing conditions

- Balances competing uses of water through efficient allocation that addresses social values, cost effectiveness and environmental benefits and costs;
- Requires the participation of all units of government and stakeholders in decision-making through a process of coordination and conflict resolution;
- Promotes water conservation, reuse, source protection and supply development to enhance water quality and quantity;
- Fosters public health, safety and community goodwill.

This definition focuses on the broad aspects of water supply. Examples can be given for other situations, including water-quality management planning, water allocation and flood control.

BALANCING DIVERSE INTERESTS

Water is a natural resource critical to the environment. Water also is an economic resource critical to society. Unfortunately, people who champion water as an economic resource essential to society commonly see themselves in direct opposition to those who champion water as a critical component of the environment that must be conserved and protected. This creates a difficult situation for those who must manage the available water resources to best satisfy all of the diverse interests that depend on water.

It is, of course, impossible to successfully manage water resources solely from either of the two extremes. If public policy exclusively protected ecological interests at the expense of economic interests, then it could threaten society's ability to meet the basic need all people have for water. From an ecological perspective, this could weaken government and diminish the ability to protect the environment, whereas from an economic standpoint, jobs and essential services such as housing and communications could be lost.

Likewise, if water management policies focus entirely on the economy, then the environment often is harmed. A diminished environment affects the public health, the quality of life and ultimately the ability to survive on planet Earth. Clearly, society is as dependent on a healthy environment as it is on a healthy economy.

LEGISLATION

In many Western Hemisphere countries, water resources are managed through legislation and regulations. In the United States, elected officials in Congress and the state legislatures set the policies that prioritize where water can or cannot be used and how much water must remain to function in the natural environment.

These laws are then implemented through regulatory agencies at the federal, state and local levels of government. One example of this legislation is the federal Clean Water Act, which requires that the quality of surface-water bodies, such as rivers and lakes, be kept at a level that will assure the maintenance of both environmental and economic functions.

Public Opinion

The effectiveness of these policies to meet both environmental and economic interests is dependent, in part, on how well the laws are written and how effectively they are enforced. However, public opinion strongly affects the type of policies that get enacted. For example, in the 1930s, the United States was emerging from a severe economic depression, so the growth and protection of the economy were paramount in the minds of the public and policymakers alike. As a result, the national policy from roughly 1930 to 1970 was to use the water of major rivers to provide jobs and to allow cities to grow in order to expand the economy. This was often accomplished through the federal government's dam-building projects for hydroelectric power and water supply, particularly for irrigation.

Shifting Sentiment

The dam-building policy was beneficial to the economy, but by 1970 national policy and public sentiment had shifted towards the environmental health of rivers and lakes, many of which had been harmed by the dam projects. One outgrowth of this environmental pendulum shift was the Endangered Species Act (ESA) passed by the U.S. Congress in 1973. The ESA is an example of legislation that fundamentally changed water policy without directly addressing water issues. On its face, the law would not seem to be about water policy. When applied to fish species, however, it places requirements on water resource management to meet certain stream flow levels in order to protect a given species. Though this is clearly a good policy, it often requires extraordinary efforts in the regional management of water to ensure that the water demands of the society are met without further endangering the listed species. These efforts can be very costly. A balance must be found whereby the environment is protected to maintain the function of the rivers and lakes while still meeting the water demands of the community.

WATER MANAGEMENT

Even without the demands of environmental laws, minimizing impacts on streams and rivers while maintaining water supplies for cities and industry is a complex task because of natural constraints on water availability. Areas of the United States like the arid Southwest have limited amounts of available water such that any additional use has the potential to significantly harm natural systems. However, water resource management is difficult even in areas with substantial annual rainfall. Here the problem is often one of timing because water may be abundant only in the winter months when water demand is typically low. Conversely, satisfying water demands for irrigation, meeting minimum-streamflow regulations and providing water to cities is typically far greater in the summer months.

All of this complicates the job of the water resource manager. As with any problem where the timing of supply is not well matched with the timing of demand, the best solution is effective storage. For example, if the annual floods of rivers like the Mississippi River could be captured and stored for use later in the year, drought would be less of a problem. In this way, sufficient water is often available to meet both ecological and economic needs provided the problem is considered regionally and over a long enough period of time.

The world is now in an era where neither the economy nor the environment can absorb much negative impact from the mismanagement of water resources. Regional watershed management may help shape policy to store water from the high-flow periods of watersheds and use that water to maintain flows for fish and wildlife requirements and meet the demands of the community during low-flow periods. Everyone should continue to strive for balanced policies that allow for the use of water without inordinate harm to either the environment or the economy. The need to balance the various and diverse interests that depend on water is more essential and more complex than ever.

3

Irrigation and Artificial Application of Water

Irrigation may be defined as the science of artificial application of water to the land or soil. It is used to assist in the growing of agricultural crops, maintenance of landscapes, and revegetation of disturbed soils in dry areas and during periods of inadequate rainfall. Additionally, irrigation also has a few other uses in crop production, which include protecting plants against frost, suppressing weed growing in grain fields and helping in preventing soil consolidation.

In contrast, agriculture that relies only on direct rainfall is referred to as rain-fed or dryland farming. Irrigation systems are also used for dust suppression, disposal of sewage, and in mining. Irrigation is often studied together with drainage, which is the natural or artificial removal of surface and sub-surface water from a given area. Irrigation is also a term used in medical/dental fields to refer to flushing and washing out anything with water or another liquid.

RESOURCE PLANNING OF IRRIGATION

Integrated resource planning is a somewhat more encompassing term than least-cost utility planning, although the two are consistent and can be used interchangeably for many analytical purposes. In fact, the term "least-cost integrated resource planning" sometimes is used. The concept of integrated planning evolved in part to address the potential misconceptions and complexities arising from use of the term "least cost" as well as any unjustified bias against supply-side solutions.

COMPREHENSIVE AND PARTICIPATORY

Integrated resource planning encompasses the concept of least-cost planning, which emphasizes balancing supply and demand management considerations and identifying feasible planning alternatives that meet the test of least cost without unduly sacrificing other policy goals. Integrated resource planning is a more comprehensive evaluation system that goes further to emphasize the construction of various planning scenarios in which key variables and assumptions can be altered. These scenarios can be used to help utilities incorporate uncertainties, environmental externalities and community needs into decision-making.

Integrated resource planning also emphasizes the importance of establishing a more open and participatory decision-making process and coordinating the many water institutions that covern water resources. Thus, IRP encourages the development of new institutional roles in addition to new analytical tools. It also promotes consensus building and alternative dispute resolution over conflict and litigation. Importantly, planning does not preclude the development and use of markets or market-like mechanisms (such as competitive bidding), which in fact may be essential for the purpose of identifying least-cost opportunities.

Integrative Assessments

Like least-cost planning, IRP explicitly recognizes that demand management can be a cost-effective and viable resource option. In a somewhat broadened sense, IRP recognizes that demand management can help achieve multiple policy goals (such as cost control and pollution prevention). Advocates of IRP have long argued that better planning methods are needed to account for environmental and social externalities associated with expanding utility capacity. More recently, analysts have recognized that IRP can help utilities deal with uncertainties and risks as well.

The generalized concept of IRP also can be used to address short- and long-term community needs that span from environmental protection to economic development. Prescreening resource options and the construction of alternative planning scenarios can be used to evaluate the implications of a given resource mix on the utility, the environment and the community. In addition to integrating resource options, IRP also seeks integration along temporal (short- and long-term) and spatial (local and regional) dimensions.

Internal and External Linkages

Integrated resource planning clearly entails new roles and responsibilities for water-supply utilities. Integration means that environmental, engineering, public health, financial, rate-making, social and economic considerations all feed into the planning process.

Planning data and information are linked internally to the other management activities of the water utility (physical facilities management, financial management, environmental management, research and development, economic development and public involvement). Integrated planning also links water utility planning with external planning processes (planning by other water, wastewater and energy utilities; local and regional planning; river basin planning; and statewide, interstate and federal water planning and policy). Some of these relationships are formal (as in permit processes involving state water resource or drinking-water regulators), while others are less so (as in the use of regional water planning data by the utility to develop forecasts). For the water sector, a particularly important issue is the relationship of water utility planning to the activities of various government agencies whose policies may constrain utility planning choices. Clearly, better coordination mechanisms could be established among the three or more state agencies that regulate water systems from the public health, natural resource and economic perspectives.

If IRP does nothing else but facilitate coordination and improvement in the institutional structures governing water resources, it will be worth the effort. In the 1990s, total water management emerged as a potentially salient concept for water and wastewater utilities. Total water management reflects the philosophy that water resources should be managed for the greatest good of people and the environment with opportunities for participation in water policy by all segments of society.

While not strictly a planning model, total water management seems to encompass the basic principles of IRP. It also has considerable symbolic connotations. According to a white paper by the American Water Works Association, "Total water management recognizes the paradigm shift from considering water available in unlimited quantities to understanding water supply as a limited resource." Total water management seeks to inspire the water industry to embrace such ideas as sustainability, stewardship, unified water resource policies, watershed and ecosystem management, water conservation and the importance of public and political support for water management decisions.

Total water management recognizes that water resources are a part of numerous complex systems, both natural and social. Advocates of integrates resources planning (encompassing, for example, water, energy and land-use planning) make a similar point. These perspectives present numerous intellectual, analytical and evaluative challenges. The term "total water management" and its manifestations are more normative and prescriptive than the seemingly value-neutral realm of rational analysis. The explicit consideration of values in planning can be uncomfortable, particularly for members of the many science-based disciplines involved in water resource issues. But in making policy choices, trade-offs among competing values are inevitable. A paradigm that allows for a dialogue about how to make these trade-offs should be welcomed by the water industry.

PLANNING IN PRACTICE

The practice of planning requires attention to the potential barriers to success.

Three key areas of concern are:

1. Access to analytical tools and adequate information,

2. The level of commitment of utilities and regulators to considering and pursuing new options
3. The consistency of approaches and methods within the real-world context of existing utility and regulatory practices.

If these issues are relevant to least-cost planning for energy utilities, they are as much or more applicable to the case of water.

Information Base

For water systems, information resources vary substantially and the need to develop data processing and analytical capabilities is clear. Attention should be paid to the design and implementation of planning strategies that minimize the effects of inadequate information.

Participants

The second barrier involves the attitudes and dispositions of those involved in resource planning. In part, success in IRP will depend on whether the prevailing corporate culture accepts conservation and related concepts as legitimate utility goals.

Realistic Approaches

Finally, successful IRP requires good practice as well as good theory. Water utility managers need practical methods for dealing with new forms of uncertainty, such as the revenue uncertainty associated with conservation.

WATER RESOURCES

Water resources are sources of water that are useful or potentially useful to humans. Uses of water include agricultural, industrial, household, recreational and environmental activities. Virtually all of these human uses require fresh water. 97 per cent of water on the Earth is salt water, leaving only 3 per cent as fresh water of which slightly over two thirds is frozen in glaciers and polar ice caps. The remaining unfrozen fresh water is mainly found as groundwater, with only a small fraction present above ground or in the air.

Fresh water is a renewable resource, yet the world's supply of clean, fresh water is steadily decreasing. Water demand already exceeds supply in many parts of the world and as the world population continues to rise, so too does the water demand. Awareness of the global importance of preserving water for ecosystem services has only recently emerged as, during the 20th century, more than half the world's wetlands have been lost along with their valuable environmental services.

Biodiversity-rich freshwater ecosystems are currently declining faster than marine or land ecosystems. The framework for allocating water resources to water users is known as water rights.

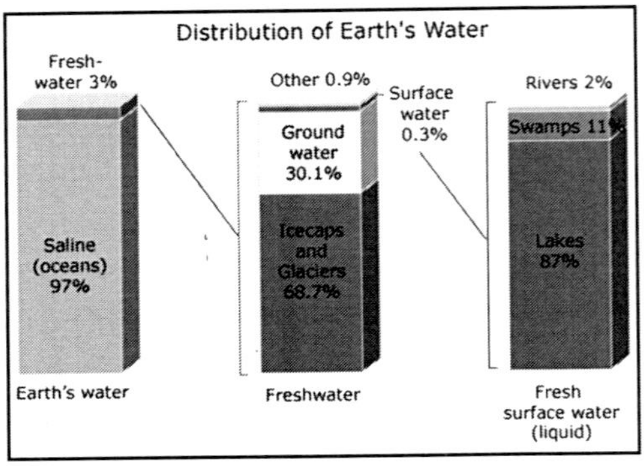

Fig. A Graphical Distribution of the Locations of Water on Earth.

SOURCES OF FRESH WATER AND SURFACE WATER

Surface water is water in a river, lake or fresh water wetland. Surface water is naturally replenished by precipitation and naturally lost through discharge to the oceans, evaporation, and sub-surface. Although the only natural input to any surface water system is precipitation within its watershed, the total quantity of water in that system at any given time is also dependent on many other factors. These factors include storage capacity in lakes, wetlands and artificial reservoirs, the permeability of the soil beneath these storage bodies, the runoff characteristics of the land in the watershed, the timing of the precipitation and local evaporation rates. All of these factors also affect the proportions of water lost.

Human activities can have a large and sometimes devastating impact on these factors. Humans often increase storage capacity by constructing reservoirs and decrease it by draining wetlands. Humans often increase runoff quantities and velocities by paving areas and channelizing stream flow. The total quantity of water available at any given time is an important consideration. Some human water users have an intermittent need for water. For example, many farms require large quantities of water in the spring, and no water at all in the winter. To supply such a farm with water, a surface water system may require a large storage capacity to collect water throughout the year and release it in a short period of time. Other users have a continuous need for water, such as a power plant that requires water for cooling. To supply such a power plant with water, a surface water system only needs enough storage capacity to fill in when average stream flow is below the power plant's need. Nevertheless, over the long term the average rate of precipitation within a watershed is the upper bound for average consumption of natural surface water from that watershed.

Natural surface water can be augmented by importing surface water from another watershed through a canal or pipeline. It can also be artificially augmented from any of the other sources listed here, however in practice the quantities are negligible. Humans can also cause surface water to be "lost" (*i.e.* become unusable) through pollution. Brazil is the country estimated to have the largest supply of fresh water in the world, followed by Russia and Canada.

UNDER RIVER FLOW

Throughout the course of the river, the total volume of water transported downstream will often be a combination of the visible free water flow together with a substantial contribution flowing through sub-surface rocks and gravels that underlie the river and its floodplain called the hyporheic zone. For many rivers in large valleys, this unseen component of flow may greatly exceed the visible flow.

The hyporheic zone often forms a dynamic interface between surface water and true ground-water receiving water from the ground water when aquifers are fully charged and contributing water to ground-water when ground waters are depleted. This is especially significant in karst areas where pot-holes and underground rivers are common.

SUB-SURFACE WATER, OR GROUND WATER

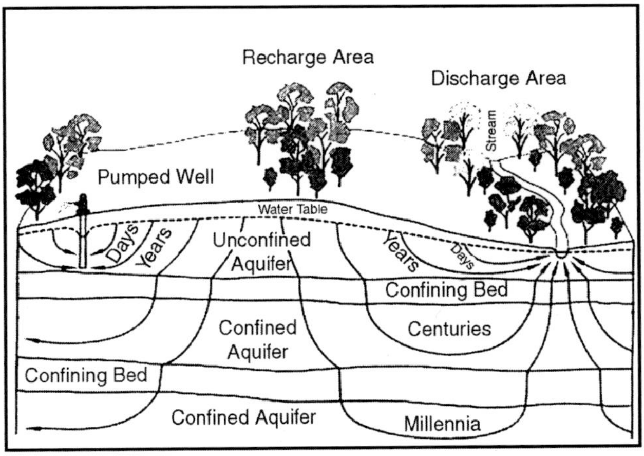

Fig. Sub-Surface Water Travel Time.

Sub-surface water, or groundwater, is fresh water located in the pore space of soil and rocks. It is also water that is flowing within aquifers below the water table. Sometimes it is useful to make a distinction between sub-surface water that is closely associated with surface water and deep sub-surface water in an aquifer (sometimes called "fossil water").

Sub-surface water can be thought of in the same terms as surface water: inputs, outputs and storage. The critical difference is that due to its slow rate of turnover, sub-surface water storage is generally much larger compared to inputs than it is for surface water. This difference makes it easy for humans to use sub-surface water unsustainably for a long time without severe consequences. Nevertheless, over the long term the average rate above a subsurface water source is the upper bound for average consumption of water from that source. The natural input to sub-surface water is seepage from surface water. The natural outputs from sub-surface water are springs and seepage to the oceans.

If the surface water source is also subject to substantial evaporation, a sub-surface water source may become saline. This situation can occur naturally under endorheic bodies of water, or artificially under irrigated farmland. In coastal areas, human use of a sub-surface water source may cause the direction of ocean to reverse which can also cause soil salinization. Humans can also cause sub-surface water to be "lost" (*i.e.* become unusable) through pollution. Humans can increase the input to a sub-surface water source by building reservoirs or detention ponds.

Desalination

Desalination is an artificial process by which saline water (generally sea water) is converted to fresh water. The most common desalination processes are distillation and reverse osmosis.

Desalination is currently expensive compared to most alternative sources of water, and only a very small fraction of total human use is satisfied by desalination. It is only economically practical for high-valued uses (such as household and industrial uses) in arid areas. The most extensive use is in the Persian Gulf.

Frozen Water

Several schemes have been proposed to make use of icebergs as a water source, however to date this has only been done for novelty purposes. Glacier runoff is considered to be surface water.

Uses of fresh water

Uses of fresh water can be categorized as consumptive and non-consumptive (sometimes called "renewable"). A use of water is consumptive if that water is not immediately available for another use. Losses to sub-surface seepage and evaporation are considered consumptive, as is water incorporated into a product (such as farm produce). Water that can be treated and returned as surface water, such as sewage, is generally considered non-consumptive if that water can be put to additional use.

Agricultural

It is estimated that 69 per cent of worldwide water use is for irrigation, with 15-35 per cent of irrigation withdrawals being unsustainable. In some areas of the world irrigation is necessary to grow any crop at all, in other areas it permits more profitable crops to be grown or enhances crop yield. Various irrigation methods involve different trade-offs between crop yield, water consumption and capital cost of equipment and structures. Irrigation methods such as most furrow and overhead sprinkler irrigation are usually less expensive but also less efficient, because much of the water evaporates or runs off.

More efficient irrigation methods include drip or trickle irrigation, surge irrigation, and some types of sprinkler systems where the sprinklers are operated near ground level. These types of systems, while more expensive, can minimize runoff and evaporation. Any system that is improperly managed can be wasteful. Another trade-off that is often insufficiently considered is salinization of sub-surface water.

Aquaculture is a small but growing agricultural use of water. Freshwater commercial fisheries may also be considered as agricultural uses of water, but have generally been assigned a lower priority than irrigation. As global populations grow, and as demand for food increases in a world with a fixed water supply, there are efforts underway to learn how to produce more food with less water, through improvements in irrigation methods and technologies, agricultural water management, crop types, and water monitoring.

Industrial

It is estimated that 15 per cent of worldwide water use is industrial. Major industrial users include power plants, which use water for cooling or as a power source (*i.e.* hydroelectric plants), ore and oil refineries, which use water in chemical processes, and manufacturing plants, which use water as a solvent.

The portion of industrial water usage that is consumptive varies widely, but as a whole is lower than agricultural use. Water is used in power generation. Hydroelectricity is electricity obtained from hydropower. Hydroelectric power comes from water driving a water turbine connected to a generator. Hydroelectricity is a low-cost, non-polluting, renewable energy source.

The energy is supplied by the sun. Heat from the sun evaporates water, which condenses as rain in higher altitudes, from where it flows down. Three Gorges Dam is the largest hydro-electric power station Pressurized water is used in water blasting and water jet cutters. Also, very high pressure water guns are used for precise cupting. It works very well, is relatively safe, and is not harmful to the environment. It is also used in the cooling of iachinery to prevent over-heating, or prevent saw blades from over-heating. Water is also used in many industrial processes and machines, such as the steam turbine and heat exchanger, in addition to its use as a chemical solvent.

Discharge of untreated water from industrial uses is pollution. Pollution includes discharged solutes (chemical pollution) and discharged coolant water (thermal pollution). Industry requires pure water for many applications and utilizes a variety of purification techniques both in water supply and discharge.

Household

It is estimated that 15 per cent of worldwide water use is for household purposes. These include drinking water, bathing, cooking, sanitation, and gardening. Basic household water requirements have been estimated by Peter Gleick at around 50 litres per person per day, excluding water for gardens. Drinking water is water that is of sufficiently high quality so that it can be consumed or used without risk of immediate or long term harm. Such water is commonly called potable water.

In most developed countries, the water supplied to households, commerce and industry is all of drinking water standard even though only a very small proportion is actually consumed or used in food preparation.

Recreation

Recreational water use is usually a very small but growing percentage of total water use. Recreational water use is mostly tied to reservoirs. If a reservoir is kept fuller than it would otherwise be for recreation, then the water retained could be categorized as recreational usage. Release of water from a few reservoirs is also timed to enhance whitewater boating, which also could be considered a recreational usage. Other examples are anglers, water skiers, nature enthusiasts and swimmers.

Recreational usage is usually non-consumptive. Golf courses are often targeted as using excessive amounts of water, especially in drier regions. It is, however, unclear whether recreational irrigation (which would include private gardens) has a noticeable effect on water resources. This is largely due to the unavailability of reliable data. Some governments, including the Californian Government, have labelled golf course usage as agricultural in order to deflect environmentalists' charges of wasting water. However, using the above figures as a basis, the actual statistical effect of this reassignment is close to zero.

Additionally, recreational usage may reduce the availability of water for other users at specific times and places. For example, water retained in a reservoir to allow boating in the late summer is not available to farmers during the spring planting season. Water released for whitewater rafting may not be available for hydroelectric generation during the time of peak electrical demand.

TECHNIQUES OF IRRIGATION

Various types of irrigation techniques differ in how the water obtained from the source is distributed within the field. In general, the goal is to supply the entire field uniformly with water, so that each plant has the amount of water it needs, neither too much nor too little.The modern methods are efficient enough to achieve this goal.

TECHNICAL DESCRIPTION

In Bangladesh, two types of agricultural water storages are observed. One type is used to store water on plots which are slightly inclined; water accumulates in the low lying portions of the fields or in a low lying areas nearby. Water stored using this method is used for both land preparation for aman paddy cropping and during the growing/milking stage of paddy farming when the water requirement is most critical. In Bangladesh, *aman* is transplanted in June/July and harvested in October/November. The other type of water storage involves the storage of rainwater in ditches or depressions located outside bunds or in land depressions beside road embankments. Water is carried from these depression to the fields in buckets, pitchers or traditional dhoon whenever it is needed. In some cases, if the waterbody is large enough, water may be used to irrigate a rabi crop or for fishing during the period of monsoonal inundation. In some upland areas of Bangladesh (North Central zone), rainwater is stored in low lying plots between two hills for use in times of necessary.

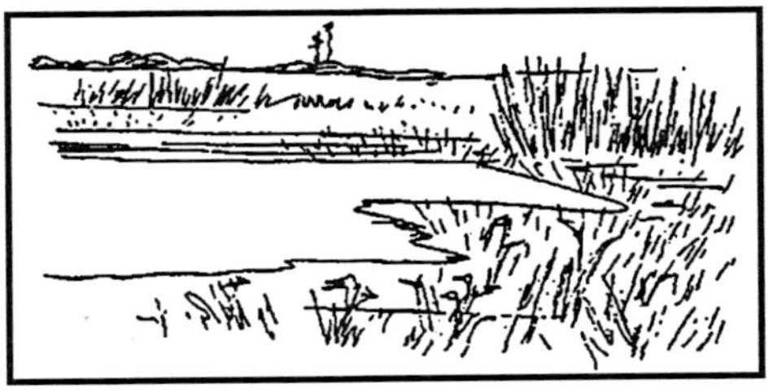

Fig. Stagnant Water on the Lower Slope of a Plot.

Rainwater stored using this technology may be transferred between fields by overflowing successive plots and may be collected in the lowest fields in a given area. In the West-Cental region, rainwater is harvested from lands situated at higher elevations and conveyed to storage ponds through culverts. In this region, because farm lands are situated at lower elevations than the storage ponds, stored rainwater may also be used for irrigating farm fields during the dry season. The specific location of the catchment area and its elevation relative to the fields to be irrigated are important factors in determining the potential for using these types of rainwater harvesting technologies.

Fig. Irrigation of Crops with Rainwater in Upland/Hilly Areas..

Extent of Use

Rainwater storage in low lying portions of farms or in neighbouring plots is possible only when the locations and elevations of the plots are suited to such storage. No detailed data on the extent of use of this technology were identified, but the field survey suggested that, if such opportunities exist, farmers made use of these technologies. In upland areas, nearly all farms having access to rainwater stored in depressions within hills made use of such water for irrigating aman paddies.

In saline areas, this practice was observed on lands located within polders or embankments erected to obstruct the intrusion of saline water. In these areas, there is a conflict between aquacultural (saline water-based shrimp culture) and agricultural (freshwater-based crop) production, which has led to violent confrontations. During the field visit, the murder of a local womens' leader, named Korunamoey Sardar and aged around 40 years, allegedly by shrimp growers has caused continued deep resentment among the local farmers and has strengthened the resolve of the local people to continue fresh water cultivation within the polder in defiance of the threats by shrimp cultivators.

Fig. Irrigation of Crops with Rainwater Using Traditional Methods.

Operation and Maintenance

The use of natural depressions does not eliminate the maintenance requirements. Repair work, such as repairing bunds, maintaining slopes, etc. involve about two person-days of family labour per year at an equivalent cost of \$2.50, based upon a rate of \$1.25 per day. Local agricultural equipment is used for this purpose.

Level of Involvement

This technology is implemented at the household level, although it has been implemented sometimes at the community level.

Costs

Construction costs are minimal since this system of rainwater harvesting is practised only when the land is naturally sloped, when the farmer has a plot nearby with a low enough elevation, or when there is an existing ditch or depression available where rainwater can be stored.

Effectiveness of the Technology

Data from the North Central zone of Bangladesh shows that the use of this technology increases paddy crop yields by about 32 per cent. In order to achieve similar yields as can be achieved through irrigation using rainwater harvesting techniques, farmers would have to incur additional expenses in transporting water from other sources. The cost of transporting water would be the equivalent of five to six person-days of family labour and one to two days of hired labour.

Suitability

This technology is suitable in areas where natural depressions are found and where there is adequate rainfall to produce runoff.

Advantages

The advantages of this technology are that it is a locally implemented technology, which can be used at little additional cost.

Disadvantages

The disadvantage of this technology is that rainwater cannot be stored for long time due to seepage losses. The method also lacks scientific rigour.

Cultural Acceptability

This technology is very well accepted and has no known cultural disadvantages.

Further Development of the Technology

This is a fully developed technology. However, development of criteria for sizing storage areas would make this technique more rigorous.

RAINWATER HARVESTING FOR COMMUNITY WATER SUPPLY

Technical Description

Rainwater is collected from rooftops of buildings using corrugated galvanized iron (CGI) sheets as the roofing material, a half-cut HDPE pipe gutter, and HDPE down-pipe to collect rainwater in ferrocement storage tanks. Some tanks have separate tapstands. This technology was introduced into Nepal for the first time in 1988. As a pilot project, a 20 m3 ferrocement tank was built to collect water from the roof of a middle school in the Village of Daungha. The system helped to fulfil the drinking water demand of about 300 students and teachers. The success of this community-based system, shown in Figure 13, encouraged the villagers to build more such systems, including an additional storage tank for the middle school. Two further 20 m3 storage tanks were constructed to harvest rainwater from the Village Committee Office building and Primary School building

Extent of Use

This technology is an example of a community-based rainwater harvesting system in Nepal. The example is from the Village of Daungha, in the Gulmi District in West Nepal. The interest of villagers and their active participation have helped to promote the development of such rainwater harvesting systems in Nepal, and there are currently 11 ferrocement tanks for rainwater storage in the region.

Operation and Maintenance

The following maintenance work is required in the operation of this technology:

- Regular cleaning of the rooftops and gutters: Rainwater is generally considered to be free of contamination, and, hence, it requires no treatment. Nevertheless, the roofs and gutters should be cleaned regularly to remove particulates and accumulated materials. Rainwater of the first few hours of the beginning of rainy season should not be collected in the tank, but should be used for flushing the roofs and gutters. Asbestos cement sheets and metal sheet roofing coated with lead-based paints should be avoided as they may be dangerous to health.
- Frequent cleaning of the storage tanks: The storage tank is the most expensive part of any rain water catchment system and determining the most appropriate capacity for any given locality and catchment area will critically affect both its cost and the amount of water available for supply. To ensure the longevity of the tank and quality of the water supply, regular cleaning of the tank should be carried out.
- Inspection of gutters and feeder pipes and valve chambers to detect and repair leaks: Rooftops are commonly used as catchments, even though ground catchments can provide a larger catchment areas, yield a greater volume of water to be collected, and are cheaper to construct. However, the use of ground catchments competes with agriculture for available land. Hence, in Nepal, both flat roofs, with tiles or plastered concrete leading to a floor drain, and sloped roofs are used. Sloped roofs are preferred to flat roofs because sloped roofs are accessible only for cleaning and repair purposes, and the harvested rainwater has less chance of being contaminated. Regular inspection of the drainage systems and conveyance systems minimizes water loss.

However, because this technology can be constructed and maintained locally, the Nepalese projects have been handed over to the users committees which bear the overall responsibilities for operation and maintenance of the systems. Such works as may be beyond the capacities and means of the users committees are carried out by the District Water Supply Offices.

Level of Involvement

Implementation of this technology has involved both the local communities and the government.

Costs

Based upon the experience in Nepal, the total cost of a 20 m3 water supply project, supplying 1 224 l/day, is about $24 620, or about $121 per person. Although the operation and maintenance costs of the system are negligible, the capital cost is too high for individual households and a rural community to invest independently in such a system. Thus, the cost of these community-based systems was divided between the government and community in the following ratio:

- Government - $22 560
- Village - $ 2 060.

The cost of a 20 m3 ferrocement tank is about $2 000.

Effectiveness of the Technology

A total of 183 m3 of rainwater is collected annually in the storage tanks served by 160 m2 of rooftop catchment area in the Daungha Village. Assuming that the water is used for drinking purposes only, it is estimated that about 73 m3 of water could meet the needs of the 34 residents for a year. With this system, water collected during the monsoon helps meet demand during the dry season. Of the remaining volume of water harvested, the per capita allocation of six litres per capita per day is considered to be too low to completely meet the demand for water for personal hygiene. However, the estimated 20 litres per capita per day supply for domestic use could be met from a household level 22 m3 rainfall harvesting system. Such a system would require a rooftop area of 49 m2 roof area and would provide water to an average household of seven members.

Suitability

This technology is suitable for use in areas with adequate rainfall and in villages having a cluster of roofs or large buildings.

Advantages

The main advantage of having a community-based rainwater harvesting system is the time savings accrued in fetching water. This frees up time which may be utilized for other economic activities. Women, who traditionally gather water, gain more time for child care, social activities and income generating activities.

Disadvantages

The disadvantage of this technology is its high initial cost and per capita cost in small settlements.

Further Development of the Technology

While the technology may be considered to be fully developed, in order to determine the potential rainwater supply for a given catchment, reliable rainfall data (mean annual rainfall and its distribution) are required for a period of at least 20 years. Improved rainfall distribution analysis methods could enhance the utilization of this technology.

SURFACE IRRIGATION SYSTEMS

In surface irrigation systems, water moves over and across the land by simple gravity flow in order to wet it and to infiltrate into the soil. Surface irrigation can be subdivided into furrow, *borderstrip or basin irrigation*. It is often called flood irrigation when the irrigation results in flooding or near flooding of the cultivated land.

Historically, this has been the most common method of irrigating agricultural land. Where water levels from the irrigation source permit, the levels are controlled by dikes, usually plugged by soil. This is often seen in terraced rice fields (rice paddies), where the method is used to flood or control the level of water in each distinct field. In some cases, the water is pumped, or lifted by human or animal power to the level of the land.

Localized

Localized irrigation is a system where water is distributed under low pressure through a piped network, in a pre-determined pattern, and applied as a small discharge to each plant or adjacent to it. Drip irrigation, spray or micro-sprinkler irrigation and bubbler irrigation belong to this category of irrigation methods.

Drip

Drip irrigation, also known as trickle irrigation, functions as its name suggests.In this system water falls drop by drop just at the position of roots. Water is delivered at or near the root zone of plants, drop by drop. This method can be the most water-efficient method of irrigation, if managed properly, since evaporation and run-off are minimized. In modern agriculture, drip irrigation is often combined with plastic mulch, further reducing evaporation, and is also the means of delivery of fertilizer. The process is known as *fertigation*. Deep percolation, where water moves below the root zone, can occur if a drip system is operated for too long or if the delivery rate is too high.

Drip irrigation methods range from very high-tech and computerized to low-tech and labour-intensive. Lower water pressures are usually needed than for most other types of systems, with the exception of low energy centre pivot systems and surface irrigation systems, and the system can be designed for uniformity throughout a field or for precise water delivery to individual plants in a landscape containing a mix of plant species. Ahthough it is difficult to regulate pressure on steep slopes, pressure compensating emitters are available, so the field does nop have to be level. High-tech solutions involve precisely calibrated emitters located along lines of tubing that extend from a computerized set of valves.

Sprinkler

In sprinkler or overhead irrigation, water is piped to one or more central locations within the field and distributed by overhead high-pressure sprinklers or guns. A system utilizing sprinklers, sprays, or guns mounted overhead on permanently installed risers is often referred to as a *solid-set* irrigation system. Higher pressure sprinklers that rotate are called *rotors* and are driven by a ball drive, gear drive, or impact mechanism. Rotors can be designed to rotate in a full or partial circle.

Guns are similar to rotors, except that they generally operate at very high pressures of 40 to 130 lbf/in^2 (275 to 900 kPa) and flows of 50 to 1200 US gal/min (3 to 76 L/s), usually with nozzle diameters in the range of 0.5 to 1.9 inches (10 to 50 mm). Guns are used not only for irrigation, but also for industrial applications such as dust suppression and logging. Sprinklers can also be mounted on moving platforms connected to the water source by a hose. Automatically moving wheeled systems known as *traveling sprinklers* may irrigate areas such as small farms, sports fields, parks, pastures, and cemeteries unattended.

Most of these utilize a length of polyethylene tubing wound on a steel drum. As the tubing is wound on the drum powered by the irrigation water or a small gas engine, the sprinkler is pulled across the field. When the sprinkler arrives back at the reel the system shuts off. This type of system is known to most people as a "waterreel" traveling irrigation sprinkler and they are used extensively for dust suppression, irrigation, and land application of waste water.

Other travellers use a flat rubber hose that is dragged along behind while the sprinkler platform is pulled by a cable. These cable-type travellers are definitely old technology and their use is limited in today's modern irrigation projects.

Centre Pivot

Centre pivot irrigation is a form of sprinkler irrigation consisting of several segments of pipe (usually galvanized steel or aluminum) joined together and supported by trusses, mounted on wheeled towers with sprinklers positioned along its length. The system moves in a circular pattern and is fed with water from the pivot point at the centre of the arc. These systems are found and used in all parts of the world and allow irrigation of all types of terrain. Newer systems have drop sprinkler heads as shown in the image that follows.

Most centre pivot systems now have drops hanging from a u-shaped pipe attached at the top of the pipe with sprinkler heads that are positioned a few feet (at most) above the crop, thus limiting evaporative losses. Drops can also be used with drag hoses or bubblers that deposit the water directly on the ground between crops.

Crops are often planted in a circle to conform to the centre pivot. This type of system is known as LEPA (Low Energy Precision Application). Originally, most centre pivots were water powered. These were replaced by hydraulic systems (*T-L Irrigation*) and electric motor driven systems (Reinke, Valley, Zimmatic). Many modern pivots feature GPS devices.

Lateral Move (Side Roll, Wheel Line)

A series of pipes, each with a wheel of about 1.5 m diameter permanently affixed to its midpoint and sprinklers along its length, are coupled together at one edge of a field. Water is supplied at one end using a large hose. After sufficient water has been applied, the hose is removed and the remaining assembly rotated either by hand or with a purpose-built mechanism, so that the sprinklers move 10 m across the field. The hose is reconnected.

The process is repeated until the opposite edge of the field is reached. This system is less expensive to install than a centre pivot, but much more labour intensive to operate, and it is limited in the amount of water it can carry. Most systems utilize 4 or 5-inch (130 mm) diameter aluminum pipe. One feature of a lateral move system is that it consists of sections that can be easily disconnected. They are most often used for small or oddly shaped fields, such as those found in hilly or mountainous regions, or in regions where labour is inexpensive.

Subirrigation also sometimes called *seepage irrigation* has been used for many years in field crops in areas with high water tables. It is a method of artificially raising the water table to allow the soil to be moistened from below the plants' root zone. Often those systems are located on permanent grasslands in lowlands or river valleys and combined with drainage infrastructure. A system of pumping stations, canals, weirs and gates allows it to increase or decrease the water level in a network of ditches and thereby control the water table. Sub-irrigation is also used in commercial greenhouse production, usually for potted plants. Water is delivered from below, absorbed upwards, and the excess collected for recycling. Typically, a solution of water and nutrients floods a container or flows through a trough for a short period of time, 10–20 minutes, and is then pumped back into a holding tank for reuse. Sub-irrigation in greenhouses requires fairly sophisticated, expensive equipment and management.

Advantages are water and nutrient conservation, and labour-saving through lowered system maintenance and automation. It is similar in principle and action to subsurface drip irrigation.

Manual using Buckets or Watering Cans

These systems have low requirements for infrastructure and technical equipment but need high labour inputs. Irrigation using watering cans is to be found for example in peri-urban agriculture around large cities in some African countries.

Automatic, Non-electric using Buckets and Ropes

Besides the common manual watering by bucket, an automated, natural version of this also exist. Using plain polyester ropes combined with a prepared ground mixture can be used to water plants from a vessel filled with water.

The ground mixture would need to be made depending on the plant itself, yet would mostly consist of black potting soil, vermiculite and perlite. This system would (with certain crops) allow to save expenses as it does not consume any electricity and only little water (unlike sprinklers, water timers,...). However, it may only be used with certain crops (probably mostly larger crops that do not need a humid environment; perhaps *e.g.* paprikas).

Using Water Condensed from Humid Air

In countries where at night, humid air sweeps the countryside, water can be obtained from the humid air by condensation onto cold surfaces. This is for example practiced in the vineyards at Lanzarote using stones to condense water or with various fog collectors based on canvas or foil sheets.

IRRIGATION WATER AND GROUNDWATER

Sources of irrigation water can be groundwater extracted from springs or by using wells, surface water withdrawn from rivers, lakes or reservoirs or non-conventional sources like treated wastewater, desalinated water or drainage water. A special form of irrigation using surface water is spate irrigation, also called floodwater harvesting.

In case of a flood (spate) water is diverted to normally dry river beds (wadis) using a network of dams, gates and channels and spread over large areas. The moisture stored in the soil will be used thereafter to grow crops.

Spate irrigation areas are in particular located in semi-arid or arid, mountainous regions. While floodwater harvesting belongs to the accepted irrigation methods, rainwater harvesting is usually not considered as a form of irrigation. Rainwater harvesting is the collection of run-off water from roofs or unused land and the concentration of this. Some of Ancient India's water systems were pulled by oxen.

Around 90per cent of wastewater produced globally remains untreated, causing widespread water pollution, especially in low-income countries. Increasingly, agriculture is using untreated wastewater as a source of irrigation water. Cities provide lucrative markets for fresh produce, so are attractive to farmers. However, because agriculture has to compete for increasingly scarce water resources with industry and municipal users, there is often no alternative for farmers but to use water polluted with urban waste, including sewage, directly to water their crops. There can be significant health hazards related to using water loaded with pathogens in this way, especially if people eat raw vegetables that have been irrigated with the polluted water.

The International Water Management Institute has worked in India, Pakistan, Vietnam, Ghana, Ethiopia, Mexico and other countries on various projects aimed at assessing and reducing risks of wastewater irrigation. They advocate a 'multiple-barrier' approach to wastewater use, where farmers are encouraged to adopt various risk-reducing behaviours. These include ceasing irrigation a few days before harvesting to allow pathogens to die off in the sunlight, applying water carefully so it does not contaminate leaves likely to be eaten raw, cleaning vegetables with disinfectant or allowing fecal sludge used in farming to dry before being used as a human manure. The World Health Organization has developed guidelines for safe water use.

WATER SCARCITY

Fifty years ago, the common perception was that water was an infinite resource. At that time, there were fewer than half the current number of people on the planet. People were not as wealthy as today, consumed fewer calories and ate less meat, so less water was needed to produce their food. They required a third of the volume of water we presently take from rivers. Today, the competition for water resources is much more intense. This is because there are now more than seven billion people on the planet, their consumption of water-thirsty meat and vegetables is rising, and there is increasing competition for water from industry, urbanisation and biofuel crops. To avoid a global water crisis, farmers will have to strive to increase productivity to meet growing demands for food, while industry and cities find ways to use water more efficiently.

Successful agriculture is dependent upon farmers having sufficient access to water. However, water scarcity is already a critical constraint to farming in many parts of the world. With regards to agriculture, the World Bank targets food production and water management as an increasingly global issue that is fostering a growing debate. Physical water scarcity is where there is not enough water to meet all demands, including that needed for ecosystems to function effectively. Arid regions frequently suffer from physical water scarcity. It also occurs where water seems abundant but where resources are over-committed.

This can happen where there is overdevelopment of hydraulic infrastructure, usually for irrigation. Symptoms of physical water scarcity include environmental degradation and declining groundwater. Economic scarcity, meanwhile, is caused by a lack of investment in water or insufficient human capacity to satisfy the demand for water. Symptoms of economic water scarcity include a lack of infrastructure, with people often having to fetch water from rivers for domestic and agricultural uses. Some 2.8 billion people currently live in water-scarce areas.

How an in-ground Irrigation System Works

Most commercial and residential irrigation systems are "in ground" systems, which means that everything is buried in the ground. With the pipes, sprinklers, emitters (drippers), and irrigation valves being hidden, it makes for a cleaner, more presentable landscape without garden hoses or other items having to be moved around manually. This does, however, create some drawbacks in the maintenance of a completely buried system.

Controllers, Zones, and Valves

Most irrigation systems are divided into zones. A zone is a single irrigation valve and one or a group of drippers or sprinklers that are connected by pipes or tubes. Irrigation systems are divided into zones because there is usually not enough pressure and available flow to run sprinklers for an entire yard or sports field at once.

Each zone has a solenoid valve on it that is controlled via wire by an irrigation controller. The irrigation controller is either a mechanical (now the "dinosaur" type) or electrical device that signals a zone to turn on at a specific time and keeps it on for a specified amount of time. "Smart Controller" is a recent term used to describe a controller that is capable of adjusting the watering time by itself in response to current environmental conditions. The smart controller determines current conditions by means of historic weather data for the local area, a soil moisture sensors (water potential or water content), rain sensor, or in more sophisticated systems satellite feed weather station, or a combination of these.

Emitters and Sprinklers

When a zone comes on, the water flows through the lateral lines and ultimately ends up at the irrigation emitter (drip) or sprinkler heads. Many sprinklers have pipe thread inlets on the bottom of them which allows a fitting and the pipe to be attached to them. The sprinklers are usually installed with the top of the head flush with the ground surface. When the water is pressurized, the head will pop up out of the ground and water the desired area until the valve closes and shuts off that zone. Once there is no more water pressure in the lateral line, the sprinkler head will retract back into the ground. Emitters are generally laid on the soil surface or buried a few inches to reduce evaporation losses.

Problems in Irrigation

Irrigation can lead to a number of problems:

- Competition for surface water rights.
- Depletion of underground aquifers.
- Ground subsidence (*e.g.* New Orleans, Louisiana)
- Underirrigation or irrigation giving only just enough water for the plant (*e.g.* in drip line irrigation) gives poor soil salinity control which leads to increased soil salinity with consequent build up of toxic salts on soil surface in areas with high evaporation. This requires either leaching to remove these salts and a method of drainage to carry the salts away. When using drip lines, the leaching is best done regularly at certain intervals (with only a slight excess of water), so that the salt is flushed back under the plant's roots.
- Overirrigation because of poor distribution uniformity or management wastes water, chemicals, and may lead to water pollution.
- Deep drainage (from over-irrigation) may result in rising water tables which in some instances will lead to problems of irrigation salinity requiring watertable control by some form of subsurface land drainage.
- Irrigation with saline or high-sodium water may damage soil structure owing to the formation of alkaline soil.

BALANCE OF GROUNDWATER

Deep percolation was not measured, although it may be approximated by the input - output difference if steady state conditions are assumed. This assumption may be acceptable for an annual period, but not for just the irrigated or the nonirrigated periods. The water balance in Watershed A indicated that deep percolation was negligible and that the difference between inputs and outputs was less than 4 per cent of the total input or output water. The negative value obtained in Watershed *B* suggests that there was not deep percolation and that the outputs were overestimated and/or the inputs were underestimated. Thus, in Watershed *B* the actual evapotranspiration was probably lower than the calculated PET_c, since this was estimated assuming maximum potential crop yields, and the peas–beans and sunflower crops were underirrigated. Also, a nonmeasured ground water input to the Watershed *B* study area could result from precipitation in its large dry-land watershed, which could explain its negative water balance. In any case, the difference between the inputs and outputs in Watershed *B* was less than 10 per cent of the total input or output water, indicating that the closure of the water balance was satisfactory. The better closure of the water balance in Watershed A than in Watershed *B* during the study period, but also when comparing the same time period (October 1997–September 1998), suggest that results from Watershed A were more reliable.

The estimated mass of *N* fertilizer and manure applied to Watersheds *A* and *B* varied between 166 and 221 kg ha^{-1} in the irrigated seasons and between 3 and 32 kg ha^{-1} in the nonirrigated seasons, representing more than 93 per cent (irrigated season) and 57 per cent (nonirrigated season) of the total imported mass of *N*. This is an expected result due to the low average NO_3–*N* concentrations measured in the irrigation (1.14 mg L^{-1}) and precipitation (1.94 mg L^{-1}) waters. The weekly NO_3–*N* loads paralleled the weekhy drainage flow rates in both monitoring stations, and they were linearly correlated ($P < 0.001$) so that each millimetre of drainage produced losses of 0.27 (Watershed A) and 0.25 (Watershed B) kg NO_3–*N* ha^{-1}. Thus, as found in other studies, the volume of drainage basically determined the exported mass of nitrates.

The mass of NO_3–*N* exported by Watershed A drainage waters varied between 19 and 9 kg ha^{-1} in the irrigated periods and between 5 and 4 kg ha^{-1} in the nonirrigated periods. The mass of NO_3–*N* exported by the Watershed B drainage waters was higher than that in Watershed A and similar (*i.e.*, approximately 25 kg ha^{-1}) in the irrigated and nonirrigated periods. As previously indicated, the higher NO_3–*N* loads in Watershed *B* were mainly attributed to the higher drainage flows derived from the precipitation falling in its larger watershed. Other reasons for the higher NO_3–*N* loads in Watershed *B* could be its larger relative area cropped with maize (57 per cent and 42–54 per cent of the irrigated areas in Watersheds *B* and *A*, respectively), which was the crop with the highest N fertilizer rates, and also the lower postplant application of *N* to maize through the irrigation system (40 and 70 per cent of the postplant *N* in Watersheds *B* and *A*, respectively). The appropriate splitting of *N* in the irrigation water has been proposed as an efficient and low-cost practice to reduce nitrate leaching losses in areas of low precipitation.

The higher NO_3–*N* load in Watershed *A* during the irrigation period of 1997 compared with the irrigation period of 1998 was attributed to the above-average precipitation of 314 *mm* in 1997, as compared with the precipitation of 196 *mm* in 1997, which was similar to the historical average. Since precipitation events are for the most part unpredictable, it should be recognized that some nitrate losses are unavoidable, even under a farmer's best management practices.

The mass of exported N expressed in per cent of the applied fertilizer plus manure *N* was 5 to 9 per cent for irrigation periods and 15 to 25 per cent for nonirrigation periods in Watershed *A*, and 11 per cent for the irrigation period and 733 per cent for the nonirrigation period in Watershed *B*. Considering the entire study periods for each watershed, NO_3–*N* loads represented 8.5 per cent of the applied fertilizer plus manure *N* in Watershed *A* and 21.8 per cent in Watershed *B*. Since closure of the water balance was better in Watershed *A* than in Watershed B and the hydrological limits of Watershed *B* were not restricted to the studied area, the 8.5 per cent figure found in Watershed *A* can be considered as more representative of the NO_3–*N* load losses derived from irrigated agriculture in this area.

The NO_3–*N* loads measured in the drainage waters were, in general, lower than those reported in other watersheds of the Ebro River basin with similar crops and climatic conditions but surface-irrigated. Basso (1994) reported in a three-year study mean annual NO_3–*N* losses of 35 to 59 kg ha^{-1} in the Bardenas I irrigation district, which represented 16 to 30 per cent of the applied fertilizer *N*. Causapé *et al.* (2002) found annual NO_3–*N* losses of 98 and 195 kg ha^{-1}, representing 44 and 56 per cent of the applied fertilizer *N*, in two watersheds of the Bardenas I district. Finally, Isidoro (1999) reported in a two-year study in the La Violada irrigation district (Monegros I) mean annual NO_3–*N* losses of 68 kg ha^{-1} or 23 per cent of the applied fertilizer *N*.

The drainage NO_3–*N* losses, particularly those found in Watershed *A* that can be considered as more representative for our study area, were in the low range of those reported in other studies around the world. In the semiarid conditions of west-central Nebraska (394 mm of precipitation from April to September), Klocke *et al.* (1999) found under sprinkler irrigation mean annual losses of 52 kg NO_3–*N* ha^{-1} (27 per cent of applied fertilizer *N*) for continuous maize and 91 kg NO_3–*N* ha^{-1} (105 per cent of applied fertilizer *N*) for a maize-soybean [Glycine max (*L.*) Merr.] rotation. The Management Systems Evaluation Area (MSEA) results from Iowa (nonirrigated maize and soybean) found annual losses of 4 to 66 kg NO_3–*N* ha^{-1}, representing 6 to 115 per cent of the applied fertilizer *N*. Pratt (1984) found in various irrigated areas of California annual average losses of 82 kg NO_3–*N* ha^{-1} (50 per cent of the applied fertilizer *N*).

In summary, despite the relatively high drainage NO_3–*N* concentrations found in our work, the low drainage discharges derived from high-quality irrigation management resulted in NO_3–*N* loads that were in the low range of those measured in other irrigated areas. Under the semiarid climatic conditions of the Ebro River basin, most water for crops is provided through irrigation. Consequently, irrigation management greatly influences nitrate losses in drainage waters. The high quality of irrigation water in the Monegros irrigation area allows for negligible leaching fractions. Our work indicates that the high quality of irrigation management (IPI within 100 ± 15) attained with sprinkler irrigation, coupled to the postplanting splitting of fertilizer N through the irrigation systems, resulted in relatively low annual drainage NO_3–*N* losses (8.5 per cent of the applied *N* for the more representative and well-managed watershed). Nevertheless, there is room for improvement within the studied watersheds, since:

- A substantial part of the plots (17–44 per cent of the total irrigated area, depending on years and watersheds) was overirrigated,
- Farmers did not account properly for precipitation on the scheduling of irrigation,
- Postplanting irrigation depths given to promote maize emergence could be decreased,
- A large proportion (60–70 per cent) of the sampled maize fields were overfertilized, and
- A larger fraction of fertilizer N could be split through its application with the irrigation water rather than applying it as preplant fertilizer. Taking into account these considerations, nitrate leaching losses in the drainage waters of these irrigated watersheds could be further decreased.

Although nitrogen pollution regulations are generally based on maximum allowable concentrations (MAC), the mass of nitrogen exported per unit irrigated area is important for controlling the nitrogen contamination in the receiving water bodies. Thus, even though drainage nitrate concentrations in the study areas are above the MAC, their nitrogen loads can be considered low, allowing for the attainment of a sensible compromise.

SURFACE WATER QUALITY AND IRRIGATION

United Nations' predictions of global population increase to the year 2025 require an expansion of food production of about 40-45 per cent. Irrigation agriculture, which currently comprises 17 per cent of all agricultural land yet produces 36 per cent of the world's food, will be an essential component of any strategy to increase the global food supply. Currently 75 per cent of irrigated land is located in developing countries; by the year 2000 it is estimated that 90 per cent will be in developing countries.

In addition to problems of waterlogging, desertification, salinization, erosion, etc., that affect irrigated areas, the problem of downstream degradation of water quality by salts, agrochemicals and toxic leachates is a serious environmental problem. "It is of relatively recent recognition that salinization of water resources is a major and widespread phenomenon of possibly even greater concern to the sustainability of irrigation than is that of the salinization of soils, *per se*. Indeed, only in the past few years has it become apparent that trace toxic constituents, such as Se, Mo and As in agricultural drainage waters may cause pollution problems that threaten the continuation of irrigation in some projects".

PUBLIC HEALTH IMPACTS

Polluted water is a major cause of human disease, misery and death. According to the World Health Organization (WHO), as many as 4 million children die every year as a result of diarrhoea caused by water-borne infection. The bacteria most commonly found in polluted water are coliforms excreted by humans. Surface runoff and consequently non-point source pollution contributes significantly to high level of pathogens in surface water bodies. Improperly designed rural sanitary facilities also contribute to contamination of groundwater.

Agricultural pollution is both a direct and indirect cause of human health impacts. The WHO reports that nitrogen levels in groundwater have grown in many parts of the world as a result of "intensification of farming practice". This phenomenon is well known in parts of Europe. Nitrate levels have grown in some countries to the point where more than 10 per cent of the population is exposed to nitrate levels in drinking water that are above the 10 mg/l guideline. Although WHO finds no significant links between nitrate and nitrite and human cancers, the drinking water guideline is established to prevent methaemoglobinaemia to which infants are particularly susceptible.

Although the problem is less well documented, nitrogen pollution of groundwater appears also to be a problem in developing countries. Lawrence and Kumppnarachi reported nitrate concentrations approaching 40-45 mg N/l in irrigation wells that are located close to the intensively cultivated irrigated paddy fields. The variation in NO_3-N which shows a peak in the *maha* (main) cropping season when rice growing is most intensive in Sri Lanka. Reiff (1987) irrigated agriculture, notes that water pollution is both a cause and an effec in linkages between agriculture and human health. The following health impacts which apply, in particular, to developing countries, were noted by Reiff:

- Adverse environmental modifications result in improved breeding ground for vectors of disease (*e.g.* mosquitos). There is a linkage between increase in malaria in several Latin American countries and reservoir construction. Schistosomiasis (Bilharziasis), a parasiti disease affecting more than 200 million people in 70 tropical and subtropical countries, has been demonstrated to have increased dramatically in the population following reservoir construction for irrigation and hydroelectric power production. Reiff indicates that the two groups at greatest risk of infection are farm workers dedicated to the production of rice, sugar cane and vegetables and children that bathe in infested water.
- Contamination of water supplies primarily by pesticides and fertilizers. Excessive levels of many pesticides have known health effects.
- Microbiological contamination of food crops stemming from use c

water polluted by human wastes and runoff from grazing areas and stockyards. This applies both to use of polluted water for irrigation and by direct contamination of foods by washing vegetables etc. in polluted water prior to sale. In many developing countries there is little or no treatment of municipal sewage, yet urban wastewater is increasingly being used directly or recycled from receiving waters, into irrigated agriculture. The most common diseases associated with contaminated irrigation waters are cholera, typhoid, ascariasis, amoebiasis, giardiasis and enteroinvasive *E. coli*. Crops that are most implicated with spread of these diseases are ground crops that are eaten raw such as cabbage, lettuce, strawberries, etc.

- Contamination of food crops with toxic chemicals.
- Miscellaneous related health effects, including treatment of seed by organic mercury compounds, turbidity (which inhibits the effectiveness of disinfection of water for potable use), etc.

To this list can be added factors such as the potential for hormonal disruption (endocrine disruptors) in fish, animals and humans. Hormones are produced by the body's endocrine system. Because of the critical role of hormones during early development, toxicological effects on the endocrine system often have impacts on the reproductive system (Kamrin, 1995). While pesticides such as DDT have been implicated, the field of endocrine disruption is in its infancy and data which support cause and effect are not yet conclusive. It is probably safe to conclude, however, that high levels of agricultural contaminants in food and water as are found in many developing country situations have serious implications for reproduction and human health.

AGRICULTURE AND THE ARAL SEA DISASTER

The social, economic and ecological disaster that has occurred in the Aral Sea and its drainage basin since the 1960s, is the world's largest example of how poorly planned and poorly executed agricultural practices have devastated a once productive region. Although there are many other impacts on water quality in the region, improper agricultural practice is the root cause of this disaster. Virtually all agriculture is irrigated in this arid area. The Aral Sea basin includes Southern Russia, Uzbekistan, Tadjikistan and part of Kazakhstan, Kirghiztan, Turkmenistan, Afghanistan and Iran.

Data on Agricultural Water Pollution in Developing Countries

Data on water pollution in developing countries are limited. Further, such data are mostly "aggregated", not distinguishing the relative proportion of "point" and "non-point" sources. In Thailand, the Ministry of Public Health reported the results of pollution monitoring of 32 rivers. Pesticide consumption has strongly increased in all developing countries. In India, consumption increased nearly 50-fold between 1958 and 1975. Yet the Indian consumption in 1973-74 was reported to be averaging a mere 330 g/ha, compared to 1483 g in USA and 1870 g in Europe.

According to various surveys in India and Africa, 20-50 per cent of wells contain nitrate levels greater than 50 mg/1 and in some cases as high as several hundred milligrams per litre. In the developing countries, it is usually wells in villages or close to towns that contain the highest levels, suggesting that domestic excreta are the main source, though livestock wastes are particularly important in semi-arid areas where drinking troughs are close to wells.

COMPETING PARADIGMS OF WATER

The Santa Fe River emerges from high (3,500m) mountains to the East of Santa Fe City and flows 75kms in a Westerly direction to join the Rio Grande. Two reservoirs in the mountains impound the entire flow of the river for City reservoirs, providing about half the water the city uses each year. The other half of the water comes from deep wells in and near the city, including from the Santa Fe River aquifer. Santa Fe's water policies are based on 19th Century Western US water law and can be summarized in the phrase, "First in time, first in right." Earlier claims to water trump later claims, other things being equal. The most important of these other things is putting the water to "beneficial use," which means an economically productive purpose. Omitted from the law is any consideration of the water resource itself. Neither the rights of nature in general or the rights of a river in particular, are accorded a seat at the legal table.

History of Water Use Along the Santa Fe River

Indigenous Pueblo Indian tribes were already using the Santa Fe River for irrigation when the Spanish arrived in the late 1500s. With the establishment of Santa Fe as a provincial capitol in 1610, agricultural use of water intensified. The Santa Fe River provided water to a growing network of Spanish canals (acequias) which provided the food for the growing settlement. More that 30 acequias were established, irrigating roughly 800 hectares of farmland, and diverting so much water that stretches of the river were dry during the summer months.

Based on the cultural values underlying contemporary acequia agriculture, it seems safe to suggest that the colonial Spanish viewed the river primarily as a means of cultivating a secure food supply in this semi-arid environment. The rights of nature were implicitly assumed: "The tacit, underlying premise [of acequia irrigators] is that all living creatures have a right to water". Another core belief related to water was that it should be shared: "The principle of water sharing belongs to a larger moral economy that promotes cooperative economic behaviour through inculcating the core value of respecto and gendered norms of personal comportment".

The 19th Century saw major changes in the cultural orientation towards the river. The United States annexed the territory of New Mexico in 1848. Thirty years later, the first dam was built on the river, to provide municipal water for the newly established Santa Fe Water Company. Initially, the Water Company stored less than 10 per cent of the river's flow for its customers, but by the mid 1900s, that proportion increased to nearly 50 per cent. From now on, the river's water would be the basis for continued urban expansion, rather than agricultural production. Today the river is operated with the objective of storing as much water as possible in the reservoirs. The rights to the river's water were extracted from the acequia farmers through legal maneuvers by the private water company, which later sold the rights to the City of Santa Fe. The city's rights comprise about 85 per cent of the river's average flow (which is highly variable). An additional 5 per cent is owned by the heirs of the early farmers (now used for urban gardens) and the remaining 10 per cent is unallocated, usually spilling from the reservoirs when the mountain snowpack melts in late Spring.

The reservoir dams are operated by City engineers. Water is normally released from the dams in anticipation of Spring floods, and (in a recent policy evolution) for aesthetic purposes during the summer months. During the rest of the year, the river is a dry, heavily eroded ditch. The policy of keeping the river intentionally dry earned it the designation as "America's Most Endangered River" in 2007. Partly in response to this negative publicity, the City government, which controls the reservoirs, shifted its policy to allow for summer water releases into the river provided the reservoirs are ¾ full rather than completely full.

A year-round minimum environmental flow, however, as required in other countries, is not under serious consideration. The City's policy remains one of prioritizing storage at the expense of flow. In the prevailing view of the Santa Fe's municipal government (whose view matters most, since the municipality owns the river's water rights), the primary and almost exclusive purpose of the river is to provide water for the city's municipal water supply, thereby supporting existing homes and businesses plus future expansion. The idea that the river itself has a rightful claim to some of its water lies outside the prevailing paradigm, which is based on a legal perspective. If you have a right to the water, and if you are complying with the legal requirement that the water be put to "beneficial use", there is no legal restriction on water withdrawals.

Pealing back the legal and economic veneers we can expose a layer of ethics that is otherwise obscured. The willingness to accept the City's legal entitlement as an adequate justification for dewatering the river implies consent with the principle of water as a commodity that can be freely owned and traded. The environmental ethic being expressed is that water is a natural resource that can (and should) be utilized for economically productive purposes. A corollary to the "water-as-resource" principle is that the ecological context of the water (river, lake, aquifer) lies outside the concern of resource management. In terms of environmental ethics, the municipal government of Santa Fe is not being un-ethical (violating the ethical code) but merely a-ethical (lacking an ethical code).

These two principles together explain how the city water managers, who are well educated professionals and generally aware of ecological science, can take the position that the health of the dewatered Santa Fe River is not a problem. By not subscribing to the principle that the river's health has an ethical importance, the test for determining whether river health deserves to be considered as a management will be based solely on economic considerations of managing the water for greatest economic benefit. This management objective is implied in the legal terms defining the conditions for meeting the legal requirements of water-as-property, namely that the water be applied to some "beneficial" activity. This is commonly interpreted to mean domestic water, landscape or crop irrigation, or any industrial purpose. Local water attorneys are divided in their opinion as to whether leaving water in the river for environmental benefit can qualify as a "beneficial" use under the law.

Water managers in New Mexico routinely express their professional mandate as making water available to the water rights holders, and specifically to "senior" rights holders, since under the law, junior water rights can be claimed only after the senior rights holders have received their water allocation. During a water shortage, for example, the owners of the oldest recorded water rights (based on the date that a claim was filed in the state courts) have a legal right to their full amount of water, as if it were a normal or wet year, before the more junior right holder's rights get activated. In practice, there is usually a sharing of the shortages but there is no legal requirement to share. If a state water manager were to express the objective of making water available for the health of the riparian ecosystems, he would almost certainly be removed from office for advocating a violation of state water law.

Ecological Water Management

The blatant disregard for the ecological health of the Santa Fe River which flows through the middle of a state capitol which is also one of the most popular tourist destinations in the country, has stimulated local efforts to restore a "living river". An initiative from within the city's governing council led to a comprehensive Santa Fe River Corridor Master Plan in 199 which endorsed the objective of a permanently flowing Santa Fe River, but without any stipulations for how this might be accomplished. The reason given for why this objective was important were couched in broadly utilitarian values:

- We advocate surface and groundwater management that balances human use with natural resource protection. We encourage government and civic leaders to place high priority on sustaining seasonal stream flow in the river, yielding hydrologic, recreational, aesthetic, and environmental value to the community. We are committed to safeguarding the long-term integrity of the river and the entire watershed.

Pieces of this agenda have been incorporated into City policies. The municipal government has officially endorsed the goal of a living river, including a 20 per cent allocation for river flow, but subject to completion of a new pipeline to secure an additional source of water, and even then only in normal or wet years.

Under dry conditions (less than 80 per cent of normal precipitation) the river flow would be shut off in favour of reservoir storage. The precise amount of water necessary for an effective environmental flow is a matter of debate even among environmentalists, but the City's planning provision for shutting off the river completely during moderately dry years clearly lies outside the concept of "environmental flow". Without the support of legal requirements establishing minimum flows or other ecological standards, river advocates such as the Santa Fe Watershed Association rely on persuasion and community interest.

The water management approach advocated by the Santa Fe Watershed Association include the following features:

- Commit a permanent minimum environmental flow to the river, which would vary according to reservoir levels but always be greater than zero (even during droughts when the reservoirs may be empty).
- Manage the reservoirs to release the Spring runoff over a period of months rather than uncontrolled spills;
- Provide financial incentives for rooftop rain catchment systems;
- Initiate water conservation campaign and incentives, and allocate a portion of conserved water to be returned to the river;
- Enlist the help of private well owners (who account for ca. 10 per cent of total urban water use) through buy-back arrangements to free up water for the river
- Capture storm flows from streets and parking lots and infiltrate into the shallow groundwater and the river.
- Increase groundwater pumping capacity for contingency during droughts

City officials accept the general concept that a flowing water is a good thing, but they stop short of endorsing the principle of environmental flow. They argue that there is not currently "enough" water available for the river. Only after the total supply is increased can the "new" demand of water for the river be accommodated, and then only under normal or above-normal precipitation years. As a City official explained to me, "The river has to live within its means." In the eyes of the City water managers, the river is a water "consumer" which competes with other consumptive demands: drinking water, irrigated gardens and lawns, golf courses, and new housing and office buildings. Each of these existing demands needs to be "balanced" and this balancing is the art of water management. The environmentalist demand for enough water for the river to support its ecological function is viewed as an unfair concession to the environmentalists. Instead, a balanced and "prudent" approach is required.

The Problem: Cultural Values. Why is the Santa Fe River considered a competing "consumer" with no claim to its own water, when in South Africa, a minimum water flow is legally protected as an "ecological reserve"? The problem is not lack of scientific knowledge or expertise. The problem lies in the realm of cultural values, and specifically values related to the role of nature, *i.e.*, environmental ethics. In Santa Fe, the river is considered an inanimate thing, a physical channel through which water sometimes flows. It is not the subject of ethical concern, much less religious veneration. The river is not even seen as a provider of water, which now comes from dams, wells, and pipes. Nature has ceased to have direct relevance to the issues of water management. The river has become a "community amenity" like a park or a garden. Nature has been tamed and can now be managed through technologies, and allocated according to a prudent and balanced assessment of society's competing water demands.

PLANNING AND MANAGEMENT, WATER RESOURCES

Most public utilities engage in some form of planning, although the extent and scope of planning vary greatly. Utility planning can be characterized by four general approaches: traditional supply planning, least-cost utility planning, integrated resource planning and total water management. Each iteration retains the tried-and-true methods of the preceding approach while expanding the scope of the planning horizon to include new issues and potential solutions.

TRADITIONAL SUPPLY PLANNING

Traditional planning for water utilities is not that different from traditional planning by electricity utilities, which can be characterized by its focus on utility ownership and control of all production resources (including centralstation power plants), its reliance on system and financial planning processes internal to the utility and its emphasis on the goals of minimizing electricity prices and maintaining a high level of system reliability. Risk avoidance also appears to be a paramount consideration in traditional utility planning. In the case of water, ownership and control of resources is more constrained (for example, by limits to groundwater withdrawals). However, the emphasis on utility ownership of the water delivery infrastructure (plants, pipes and so on) prevails much like the case of electricity distribution.

Shortcomings

Three principal concerns about the traditional approach have been advanced. First, forecast demand is taken as a given and virtually no attempt is made to integrate supply management and demand management options.

Second, the public-at-large, outside experts and government regulators generally have little or no involvement in traditional utility planning. Demand analysis and the assessment of supply alternatives takes place within the utility (or a single planning unit within the utility); only the final product is made available for review or regulatory approval. Often, major investment decisions are made with little or no oversight. Third, traditional planning also tends to be confined to individual utilities in virtual isolation. The decision-making process excludes parties who not only have a vested interest but also unique insights about resource options. In sum, there is a tendency for traditional planning to be narrowly focused and exclusionary. Once a utility has committed itself to a plan, little room is left for altering the chosen course.

Supply-Driven Focus

Like other types of public utilities, the prevailing planning processes undertaken by water utilities have been internally driven and dominated by supply considerations. The result has been an emphasis on maintaining reliable water supplies and, accordingly, the engineering of facilities for source development, treatment and storage and transmission and distribution of water.

Water-supply planning generally takes the form of forecasting future demand and developing and analyzing supply options to meet the projected demand level, plus a comfortable margin. The result is a disaggregated planning approach focusing only on new supply alternatives, while initiatives in the areas of demand management and conservation are consigned to separate programmes. Moreover, conservation effects often are not reflected in revised demand and revenue forecasts.

Delivery, Not Demand

Engineering considerations have always been central to water utility planning, at times to the exclusion of other perspectives. Engineering does take into account the benefits and costs of supply projects. However, the emphasis on supply options can result in additions to capacity that outpace growth in demand, which is a problem familiar to the electricity industry.Water's natural abundance in many areas may explain why the central water supply issue is that of engineering water delivery (getting water to where it is needed) rather than managing water demand. Historically, it was easy to presume that there always would be enough water to go around and that plentiful amounts would generally be so inexpensive that investments in demand management would not be cost-effective. The utility corporate culture, which understandably emphasizes selling more (not less) product, might also favour the supply side. At the very least, utilities traditionally have had little incentive to promote conservation.

LEAST-COST PLANNING

The shortcomings of traditional planning, along with significant external economic and political forces, gave rise to the current interest in least-cost planning, or integrated resource planning (IRP). Least-cost planning emerged in the context of the energy industries during the 1980s as a response to rising costs, a poor record of forecasting and growing concerns about environmental externalities. According to its proponents, planning allows regulators to be more proactive, that is, to actively affect utility decisions rather than simply react to them later on.

Least-cost planning emphasizes a balanced consideration of supply management and demand management options in identifying feasible least-cost alternatives for meeting future water needs. Compared with traditional planning, least-cost planning recognizes that water demand is malleable and that forecast demand does not have to simply be taken as a given in the planning process.Different definitions emphasize the minimization of rates, customer bills, utility revenue requirements and production (both capacity and operating) costs. The variety of available perspectives leaves open the question of whose costs are supposed to be least: existing customers, future customers, utility shareholders, or society at large.

Another unfortunate implication of the concept of least cost is that it suggests a single, optimal solution. Regulators and utility managers do not necessarily have identical views about the meaning of least-cost planning. According to one survey, "regulators tend to view least-cost planning with an emphasis on conservation, whereas utilities tend to regard least-cost planning as an integrated supply-and-demand analysis."

A Broader Context

Least-cost planning has come to be understood as the comprehensive evaluation of all supply and demand alternatives with the end result, in an attempt to minimize costs, of creating a flexible plan allowing for uncertainty and a changing economic environment. Cost minimization, resource diversity, risk management and flexibility are the hallmarks of least-cost planning.

In practical terms, least-cost planning can be characterized by the consideration of a diverse set of resource options (including conservation, load management, pricing and purchases from other producers); coordination among several departments within the utility; and the use of multiple resource selection goals, including those that address prices, costs, revenue requirements, utility financial condition, risk reduction, technological diversity, environmental quality and economic development. Least-cost planning in no way abandons sound engineering practices, but it does place them in a wider context.

It follows that least-cost planning can facilitate regulatory review of supply management and demand management practices and regulatory approval of capacity expansion projects prior to construction. Least-cost planning also can be used to educate policymakers about the critical issues in utility supply decisions and induce utilities to aggressively engage in long-term strategic planning. Least-cost planning can focus on a particular utility or take the form of a statewide assessment of all utilities in a given utility sector, the latter of which is typically performed by a state resource management agency.

Demand Versus Supply

In contrast to traditional supply planning, leastcost planning gives much weight to the distinction between demand management and supply management activities, both of which can be used to achieve reliability goals. Least-cost planning recognizes that demand is malleable rather than simply a fixed input to planning models. The demand side involves any strategy to eliminate or defer the need for an investment in new capacity by the utility, including load management, conservation and pricing strategies. The supply side involves determining the most efficient method of meeting growing demand, including investments in new capacity.

All demand management activities that decrease the demand for utility services tend to affect supply management since existing system capacity is released for other customers and other uses. That is, the freed or redirected utility capacity can be compared to that provided by more traditional means. Thus emerged the concept of "negawatts," meaning electricity "produced" through conservation and efficient use of existing electricity supply resources. A related idea is the trend to establish energy service companies, as compared with traditional electric utilities, who could market efficiency (or demand management) as well as electrical power.

Complexity, Risk and Cost

Least-cost utility and planning is complicated by the lack of familiarly with demand management, barriers to coordination with nearby utilities, the use of broad definitions of costs (such as those associated with externalities) and the inclusion of goals (such as resource protection) not directly attributable to the utility. Traditional planners tend to view least-cost planning as more risky, as least in terms of supphy reliability. Advocates of least-cost planning would counter that planning actually reduces some forms of risk, such as the revenue recovery risk associated with excess capacity. While highly pertinent to the debate, these disagreements can be overcome. One issue that merits especially careful attention, however, is that of the cost of planning itself. Certainly, the benefits of planning should outweigh the costs. This calculation itself poses a rational decision-making problem. The benefits of increased awareness and understanding and reduced ignorance and uncertainty can be substantial but are not easily quantified.

INTEGRATED RASOURCE PLANNING

Integrated resource planning is a somewhat more encompassing term than least-cost utility planning, although the two are consistent and can be used interchangeably for many analytical purposes. In fact, the term "least-cost integrated resource planning" sometimes is used. The concept of integrated planning evolved in part to address the potential misconceptions and complexities arising from use of the term "least cost" as well as any unjustified bias against supply-side solutions.

Comprehensive and Participatory

Integrated resource planning encompasses the concept of least-cost planning, which emphasizes balancing supply and demand management considerations and identifying feasible planning alternatives that meet the test of least cost without unduly sacrificing other policy goals. Integrated resource planning is a more comprehensive evaluation system that goes further to emphasize the construction of various planning scenarios in which key variables and assumptions can be altered. These scenarios can be used to help utilities incorporate uncertainties, environmental externalities and community needs into decision-making.

Integrated resource planning also emphasizes the importance of establishing a more open and participatory decision-making process and coordinating the many water institutions that govern water resources. Thus, IRP encourages the development of new institutional roles in addition to new analytical tools. It also promotes consensus building and alternative dispute resolution over conflict and litigation. Importantly, planning does not preclude the development and use of markets or market-like mechanisms (such as competitive bidding), which in fact may be essential for the purpose of identifying least-cost opportunities.

Integrative Assessments

Like least-cost planning, IRP explicitly recognizes that demand management can be a cost-effective and viable resource option. In a somewhat broadened sense, IRP recognizes that demand management can help achieve multiple policy goals (such as cost control and pollution prevention).

Advocates of IRP have long argued that better planning methods are needed to account for environmental and social externalities associated with expanding utility capacity. More recently, analysts have recognized that IRP can help utilities deal with uncertainties and risks as well.

The generalized concept of IRP also can be used to address short- and long-term community needs that span from environmental protection to economic development. Prescreening resource options and the construction of alternative planning scenarios can be used to evaluate the implications of a given resource mix on the utility, the environment and the community. In addition to integrating resource options, IRP also seeks integration along temporal (short- and long-term) and spatial (local and regional) dimensions.

Internal and External Linkages

Integrated resource planning clearly entails new roles and responsibilities for water-supply utilities. Integration means that environmental, engineering, public health, financial, rate-making, social and economic considerations all feed into the planning process. Planning data and information are linked internally to the other management activities of the water utility (physical facilities management, financial management, environmental management, research and development, economic development and public involvement).

Integrated planning also links water utility planning with external planning processes (planning by other water, wastewater and energy utilities; local and regional planning; river basin planning; and statewide, interstate and federal water planning and policy). Some of these relationships are formal (as in permit processes involving state water resource or drinking-water regulators), while others are less so (as in the use of regional water planning data by the utility to develop forecasts). For the water sector, a particularly important issue is the relationship of water utility planning to the activities of various government agencies whose policies may constrain utility planning choices. Clearly, better coordination mechanisms could be established among the three or more state agencies that regulate water systems from the public health, natural resource and economic perspectives. If IRP does nothing else but facilitate coordination and improvement in the institutional structures governing water resources, it will be worth the effort.

WATER RESOURCES AND TRADITIONAL IRRIGATION SYSTEMS

Having evolved under the influence of the Himalayan rain shadow, Zangskari cultivation is entirely irrigation-dependent. Indeed, throughout Ladakh, water is widely recognised as being one of the most significant factors constraining the size of settlements and the area under cultivation. Although a number of factors influence whether a soil in Zangskar is cultivable, proximity to a reliable water source is critical. Villages are therefore positioned in order to utilise the melt-water from streams and springs which drain neighbouring glaciers and semi-permanent snowfields. Carefully engineered channels harvest this water and transport it, sometimes over several kilometres of rugged terrain, to the area under cultivation.

Villages are reliant upon small scale gravity fed channels as Zangskar's major rivers are largely unusable. Not only would spring floods in these rivers destroy the head works of the irrigation channels, but these high energy flows have also cut deep into the landscape and thus they are generally too low to feed into gravity fed irrigation systems. Influenced by a host of variables, the availability and reliability of irrigation water that is received alters from village to village. Variables may include: the aspect and size of glaciers or snowfields; climate and weather fluctuations (especially the previous winter's snowfall); the degree of channel maintenance; topography and soil/rock type between the source and the village; and the nature and size of the area under cultivation. It is therefore no surprise that the amount of water a village receives "varies by year, season, day and even hour".

In response to this variability, and potential scarcity, the water allocation systems that exist in each village incorporate a degree of flexibility into the fixed framework which ensures equitable and efficient distribution. Operating in a rotational system, water-user groups may be organised according to field, irrigation channel or household. Members of the user groups have the right to use water for an allotted period of each rotation and this 'water time' is informally regulated by a transparent system in which villagers monitor each other. Many irrigation channels in Zangskar are centuries old and the right to use water from these channels is usually inherited. Water allocation generally favours the households that claim 'full' ancestral rights and without such a claim it is very difficult to break into these closed village systems.

This 'club' mentality to water rights effectively discourages newcomers and the further division of village water resources. Such a policy also reflects the extent to which household power, productivity and wealth is linked to the securing and protection of water rights. Not only is irrigation fundamental to life in Zangskar it is also woven into the social and political fabric of Zangskari villages. As a communal resource, individuals have to seek the consent of the village before building new fields and channels that require water from the existing irrigation network. Under the organisation of the village headman (mog dam), the maintenance of village irrigation channels, or the less frequent construction of new channels, also reflects this communal approach as traditionally the labour for these tasks is provided by each household.

Even households that are unable to offer labour contribute by supplying food and drink for those working on village projects. Aside from the annual removal of silt, traditionally constructed irrigation channels generally require little structural maintenance. The roots of grasses and herbaceous plants that grow along the banks of the channel bind the soil and help maintain channel structure. However, in very sandy soils, or where the irrigation channel crosses areas of scree, excess seepage is often a problem. Whilst many villages enjoy the relative water security of large glacial reserves, others may be reliant upon smaller snowfields and the regular accumulation of winter snowfall.

Indeed, a recent spate of poor winter snowfall and resulting summer droughts graphically illustrated the vulnerability of some of Zangskar's settlements. Villages such as Karsha, Pishu and Zangla, proved to be particularly susceptible, and these and many other settlements throughout Zangskar employed drought strategies which involved decreasing, sometimes by up to two thirds, the area under cultivation. Kumi, a village that has experienced water shortages over the last 20 years, also suffered and activity is now underway to move the entire village down to the banks of the Lungnak River in an attempt to regain water security.

SETTLEMENT, HOUSEHOLDS AND POLYANDRY

According to the Indian census Zangskar consists of only 25 villages. This figure is however somewhat misleading as it refers to administrative units rather than single settlements. For example: an administrative 'village' may consist of a large settlement of some 70 or more houses, and several smaller contiguous 'hamlets' which, despite being separated from the main settlement by large expanses of unimproved land, fall under one census entry. In reality Zangskar has over 100 individual settlements although some may only contain two or three houses.

Despite being distributed over a vast and diverse landscape, Zangskar's villages share common characteristics such as their relatively low altitude and their proximity to suitable water and cultivable soil. The favourable convergence of these variables is however rare, not only in Zangskar but across much of Ladakh. Productive land is therefore highly prized and rules of inheritance ensure that the household, complete with farming estate, is passed down in its entirety to the next generation. As Gutschow points out "the Zangskari household is the principal nexus around which the economic relations of labour, production and subsistence are arranged". Continuity of the household is of primary importance and this is traditionally facilitated in Zangskar's Buddhist community through the practice of polyandrous marriage and primogeniture.

At some point after marriage the family's eldest son becomes the head of the household and inherits the 'main house' (khang pa or khang chen) and accompanying land. The eldest son also inherits the responsibility of managing household affairs and the farming estate, often with the assistance of one or more of his brothers who may also share his wife. Aged parents move to a minor house (khang chung) where, depending on the labour requirements of the khang pa, they may also be joined by unmarried siblings. In the event that there is no reproductive brother to inherit the khang pa then the responsibility of heading the household falls to the eldest daughter who engages in a matrilocal marriage. As Crook points out whichever system of marriage is settled for, and it may simply be a monogamous arrangement, the continuity of the household and the entire farming estate is maintained within the same patrilineage.

Whilst polyandry maintains household unity it is also considered, in a more functionalist interpretation, to by a practice which inhibits population growth and therefore reduces the demands of a large population on a resource poor environment. The effects of these Buddhist practices are particularly apparent in the village of Padum where, in contrast, the rules of inheritance practised by Zangskar's Muslim minority have resulted in the equal division of farming estates between siblings.

In an environment where water and cultivable land is limited this practice eventually leads to the continued fractionalisation of land and progressively smaller plots which, as Singh points out, are not agriculturally viable. In Padum the Muslim community is now expressing concern over the increasingly small plots of land on which they have to farm. However, relatively recent developments have influenced this situation such that these concerns are no longer restricted to the Muslim community.

Polyandrous marriage and primogeniture was outlawed in Ladakh by State legislation in the early 1940s. Although the effects of this were not instantaneous, especially in remote regions such as Zangskar, a slow trend away from polyandrous marriage and primogeniture was initiated. In Zangskar's Buddhist community this shift is particularly apparent in Padum and neighbouring settlements, where the road link to Kargil and the opportunity of government employment have fuelled modernisation processes and the growth of a cash economy.

Younger (Buddhist) sons who traditionally would not qualify for inheritance rights have claimed their share of land and the result, especially in and around Padum, has been a sudden rise in the number of minor households (khang chung). The new cash economy has provided a means of overcoming any resulting shortfall in agricultural productivity or available labour.

However, once outside Padum's economic sphere of influence, village growth is still restricted by the availability of cultivable land and, more importantly, water. Although not conclusively linked, the decline of polyandry and primogeniture and the increase in monogamous nuclear households may at least in part be responsible for the population increase in Zangskar. The Revenue Office and Sub-Divisional Magistrate in Padum confirm a marked population growth over the last 30 years from 6,865 in 1973 to 12,167 in 2003. Yet, despite this rapid increase, Zangskar continues to be one of the least populated areas in Kargil District.

ADMINISTRATION, DEVELOPMENT, BUDDHISTS AND MUSLIMS

Zangskar is categorised as a tehsil, a Sub-Division and a (Community Development) Block within Kargil District, one of two Districts - the other being Leh - that constitutes the Ladakh region. Kargil District contains two tehsils, Kargil and Zangskar, and these in turn are divided into Blocks. Whilst Kargil tehsil is divided into six Blocks, Zangskar tehsil contains only one as the relatively small population does not warrant further division. As development funds are allocated on a Block level, Zangskaris often consider themselves to be at a disadvantage, especially as Zangskar tehsil constitutes around 60 per cent of the District's total land area. However, the designation of Sub-Division does provide Zangskar with an additional administration intended to boost the development of remote and 'backward' areas. As a Sub-Division Zangskar benefits from a block level administration headed by a Sub-Divisional Magistrate (SDM) and a Public Works Department (PWD) headed by an Executive Engineer.

The majority of development initiatives which are implemented in Zangskar originate at State or Central Government level. Within the scope of these government initiatives, funding applications for Block-level development are submitted by Zangskar's administration to the District administration based in Kargil. Subject to District-level approval, J & K State or Central Government is then approached for funding. If successful, funding is then released back through the system to the Zangskar Block. In theory, this administrative system represents a simple and efficient hierarchy. In practice, even government employees admit that the percolation of directives and funds through this system is often slow and haphazard. Zangskar, like the rest of Ladakh, is well integrated into both Central and J & K State rural development initiatives. However, from the perspective of the Zangskari, development activity frequently falls short of addressing such fundamental issues as health care, education, agricultural development and water security. Logistical problems associated with Zangskar's physical remoteness are undoubtedly responsible for part of this perceived neglect, however, the most frequently voiced reason, especially from Zangskar's Buddhist majority, is the division between Buddhist and Muslim communities.

There is a lengthy historical and political background to Muslim and Buddhist relations in Ladakh. However, a brief example illustrates the nature of the grievances that perpetuate what essentially is a divide based on the politicisation of religion. Prior to the bifurcation of Ladakh District in 1979 and the creation of the Buddhist majority Leh District and Muslim majority Kargil District, all State and Central Government Offices were centred in the predominantly Buddhist town of Leh. Under this Buddhist majority administration, the Muslim minority considered themselves to be disadvantaged by discriminatory policies that, even today, are considered to be the root cause of Kargil District's backwardness.

Regardless of whether the accusation is founded or not 'evidence' of this nature is continually recycled in the struggle for political representation and dominance. Thus the perpetuation of distrust, resentment and religious division is assured. Far from suggesting that the current Kargil administration is employing a policy of retaliation against the Buddhist minority in Zangskar, the stage has been set for the less than cordial relations to persist. Indeed, Zangskar's Buddhists frequently accuse Kargil's Muslim majority administration of discrimination, a lack of political representation and the withholding of development funds and resources.

This division and general distrust is intensified further by an influential Muslim minority in Zangskar, who not only constitute the majority in the village of Padum, Zangskar's administrative capital, but also command a greater portion of government positions (and salaries) in Zangskar's Sub-Division administration. Events in the summer of 2003 have served only to perpetuate this situation. Kargil District has recently been granted Autonomous Hill Development Council (AHDC) status by the Jammu and Kashmir State Government. Zangskar however has only been awarded 3 out of 26 seats on the Council and vocal protests by several (Buddhist) representatives from Zangskar have been largely ignored. In protest the Buddhist dominated Zangskar Action Committee intended to boycott the Hill Council elections in July 2003 and begin 'agitating' the Muslim administration. The implications of Zangskar's geographical and political isolation have also been discussed. Building upon this foundation this study will demonstrate how these factors are influencing the continued development of Zangskar's irrigation water through the implementation of the Watershed Development Programme.

4

Drip Irrigation Systems

INTRODUCTION

Drip irrigation is today's need because Water - nature's gift to mankind is not unlimited and free forever. World water resources are fast diminishing. The one and only answer to this problem is Jain Drip Irrigation Systems. "Jain Drip" the name which you can trust, the only manufacturer of all drip irrigation components.

After detailed study of inter-relationship among soil, water, crop, land terrain and related agro climatic conditions, Jains design a suitable and economically viable system to deliver a measured quantity of water at the root zone of each plant at regular intervals. This is to ensure that the plants do not suffer from stress or strain of less and over watering. The system installed at the farmer's field is commissioned and training imparted to the farmer, followed by regular after sales services.

The result – A totally customized, efficient and long-life system which ensures saving in water, early maturity and a bountiful harvest, season after season, year after year. Apart from all this, savings in labour and fertilizer costs.

By installing Jain Drip, you will be a member of a happy family of Jain Drip System owners.

- Based on careful study of all the relevant factors like land topography, soil, water, crop and agro-climatic conditions, we select the most suitable and scientific micro irrigation system. Jains offer you a complete system for your crop so that you reap all the benefits.
- We at Jain Irrigation do not merely sell the micro irrigation system, we provide Agronomic and Extension support, after sales services and all technical supports for getting better crop returns. And for this, we have more than 300 technocrats, engineers, agronomists, horticulturists and regional offices, as well as trained dealers, distributors all over India and abroad.
- Jains are a one stop shop for total Agricultural input needs. We

have the capability and adequate support infrastructure to take up total turn-key Agricultural Development Projects of any size within the country or abroad, irrespective of land, topography, soil, water and other Agro climatic conditions.

- Jain Micro Irrigation System is made from high quality virgin raw materials, using advanced machinery. It is durable, reliable and meets International quality standards.
- Apple, grapes, banana, sugarcane, tea, coffee, cotton, mango, teak-wood, vegetables, flowers... whatever may be your crop, we have a suitable micro irrigation system for each of them. All the system components are manufactured by us in our plant at Jalgaon, under strict quality control norms at every stage of production.
- Jains have been exporting various components of Micro Irrigation System to countries in Europe, America, Africa, South East, Middle East and Far East Asia.
- Jain Micro Irrigation System means a technology developed for farmers by a company who knows and understands the farmer and his needs for four decades..

Benefits of Jain Drip Irrigation Systems

- Has recorded increase in yield up to 230 per cent.
- Saves water up to 70 per cent compare to flood irrigation. More land can be irrigated with the water thus saved.
- Crop grows consistently, healthier and matures fast.
- Early maturity results in higher and faster returns on investment.
- Fertilizer use efficiency increases by 30 per cent.
- Cost of fertilizers, inter-culturing and labour use gets reduced.
- Fertilizer and Chemical Treatment can be given through Micro Irrigation System itself.
- Undulating terrains, Saline, Water logged, Sandy & Hilly lands can also be brought under productive cultivation.

Model Design

Drip irrigation system delivers water to the crop using a network of mainlines, sub-mains and lateral lines with emission points spaced along their lengths. Each dripper/emitter, orifice supplies a measured, precisely controlled uniform application of water, nutrients, and other required growth substances directly into the root zone of the plant. Water and nutrients enter the soil from the emitters, moving into the root zone of the plants through the combined forces of gravity and capillary. In this way, the plant's withdrawal of moisture and nutrients are replenished almost immediately, ensuring that the plant never suffers from water stress, thus enhancing quality, its ability to achieve optimum growth and high yield.

POTENTIAL OF DRIP IRRIGATION

The potential of drip irrigation in improving agricultural productivity was identified long back.

Drip irrigation is the term used for describe the method of irrigation which is characterized by following features.

- Water is applied at a low rate. It is let out drop by drop at the spot of application.
- A given quantity of water is applied over a long period of time.
- Water is applied at frequent intervals.
- Water is applied directly to the root zone of plants. This minimizes the water loss.
- Water is applied via a low pressure delivery system. Usually gravitational force is used for drip irrigation.
- Since the crops are irrigated throughout their growth periods, more water could be saved by dripping irrigation.

Water management in drip irrigation is carried out through micro irrigation techniques such as surface drip and sub surface drip irrigations. Micro irrigations may be defined as a method of irrigation which is used to provide equal and even amount of water directly to the plant in a controlled manner. Surface drip and sub surface drip applications are the two types of drip irrigation. They are briefly explained as follows.

Drip Systems and Components

Drip irrigation differs from conventional (furrow or sprinkler) by providing water to a plant in a localized fashion, only at the plant root zone. Water is applied frequently, slowly and at low pressures. The field layout for a crop row spacing of one meter is detailed. The main components of a drip irrigation system is explained here.

Main Components

A drip irrigation system consists of a main line, submains, laterals and emitters. The main line delivers water to submains and the submains into the laterals. The emitters which are attached to laterals distribute water for irrigation. The mains, submains and laterals are usually made of black PVC (poly Vinyl Chloride) tubing. The emitters are also made of PVC material though other materials are also used. PVC material is preferred for drip system as it can withstand saline irrigation water. It is also not affected by chemical fertilizers. Emitters may have porous walls or twin walls and is a more complex mechanical unit. Lateral lines are about 3/8 to 3/4 inch in diameter, and the main lines are of polythene 2 to 4 inches in diameter. Emitters may lie on surface or buried under the soil as per design. The systems operates either daily or every other day, during most of the growing season. Duration of application ranges from 1 to 16 hours and generally is not continuous.

Ancillary Components

These include a valve, pressure regulator, filters, pressure gauge etc. The head of the system consists of a pump to lift the water and produce the desired pressure and distribute the water through nozzles or emitters. There are in general three main connections in a drip irrigation system, namely, the main-sub main connections and sub-main lateral connections and the lateral and microtubes.

The main sub-main connection is usually a "Tee" either thread or slip. A sub-main is used to deliver water into laterals and also used as a controller so that the field can be irrigated separately under a desired water pressure at any selected time. All the specific items used for control are installed in the sub- main which is called a sub-main manifold.

Pump and Prime Mover

The pressure necessary to force water through the components of the systems, including the filter unit, main line, lateral lines and the drip nozzles, is obtained by a pump of suitable capacity. Volute centrifugal pumps operated by engines or electric motors are commonly used.

The water pressure developed by the pump should be sufficient to maintain the desired pressure at the laterals. The laterals may be designed to operate under pressure as low as 0.15 to 0.2 Kg/cm^2 and as to 1.75 Kg/cm^2. A pressure drop of 0.5 to 1.0 Kg/cm^2 may be anticipated in the head of the drip system, including the filter. There is a further drop of pressure in the laterals and the emitters. The water coming out of the emitters is almost at atmospheric pressure.

Filter Unit

A filter unit which cleans the suspended impurities in the irrigation water so as to prevent blockage of holes and passages of microtubes is an essentia part of the drip irrigation system. The effective performance of drip nozzles lies in the effective performance of the filter unit. The filter system may consis of valves and a pressure gauge for regulation and control.

Micro Tubes

The micro tubes or 'emitters' or 'drippers' or drip nozzles are provided a a regular intervals on the laterals. They allow the water to emit at very lov rates, usually in trickles.

There are three types of trickles.

- Water seeps out continuously along the lateral lines.
- Water sprays or drips from an emitter connected or inserted on th laterals.
- Water sprays or drips from holes punched in the laterals.

The discharge rate of emitters usually range from 2 to 10 litres per hou

Drip System Design

The following general information are required for designing a drip irrigation system.

Water Source

The source of water is usually a well or a tank storing rainfall runoff water. If it is a well, information such as the level of water depth, recharging capacity, etc. is required for designing the system. If it is a tank, information on reservoir capacity, evaporative loss etc. are required to fabricate a drip irrigation system.

Types of Crops

Different crops require different plant to plant and raw to raw spacings and water requirements. The general layout of the system and especially the emitters will depend on the type of crops sown.

Topographic Condition

It is necessary to know the general level slope, land levelling and other level details to determine the size and location of main and sub-main lines.

Soils

The infiltration rate, water holding capacity, texture, structure and the bulk density of the soil must be known for designing.

Climatic Records

These will show when and how often an irrigation is needed during various seasons of the year.

Basic Hydraulics

A drip irrigation system is made by a combination of different sizes of plastic pipes which are usually considered as smooth pipes. The variation of discharge from emitters along a lateral line is a function of the total length of the laterals and inlet pressure emitting spacing, and total flow rate. This creates a design problem to select the right combination of length and pressure in order to achieve an acceptable and non- uniform pattern of irrigation. Hydraulically the pressure variation along the lateral line will cause an emitter-flow-variation and a pressure variation along a sub-main will cause a lateral line flow variation.

Surface Drip Irrigation

Surface drip irrigation is provided with a Lateral Drip Poly Ethylene (LPE) or Life Lateral Drip Poly Ethylene (LLDPE). Further they are fitted with microtubes which are also called drippers, emitters and drip nozzles as per design. The daily requirement of water is supplied by means of lateral pipes micro tube network on the surface of the soil. From these lateral pipes and micro tubes it is easy to maintain the drip irrigation system as the clogging or any other defects can be located and rectified immediately. The system can be shifted whenever and to wherever it is required with minimum damage. The cost of installation is also reduced significantly. At harvesting special care has to be taken to avoid damages to lateral pipes and micro tubes. The approximate cost of 10000 to 12000 per acre depends mostly on design characteristics, operation and maintenance. The network of drip irrigation consists of lateral pipes fitted with microtubes and running in between rows of the crops, main line fitted with pressure computing self flushing dripper valves at suitable distances. All the lateral tubes are connected with a controlled water pumping mechanism, which gradually and slowly pumps water which is made available by microtubes to the soil at regular intervals.

Sub-Surface Drip System

The LLDPE (Life Lateral Drip Poly Ethylene) lateral lines are buried in trenches at the depth of 15-25 cms and water is applied in the sub soil zone. For long duration crops the depth of trench may be about 30-45 cms. They may be strips of pipes provided with holes or twin-wail-laterals under a specific pressure compensating system maintaining uniform discharge throughout the lateral pipe. There are several prototypes of twin wall systems. The trenches are to be made at the appropriate depth manually in the fields. The twin wall laterals are laid in the trenches with openings upwards and trenches are closed with soil. The other equipments ego sand filters screen filters, pressure gauges pressure valves, water meter, pumping mechanism etc. are connected as per design.

Concepts of Drip Irrigation

Drip irrigation has the greatest potential in saving water and improving the crop yields. Drip irrigation is the frequent slow application of water at low pressure through emitters located at selected points along the delivery pipe lines. The typical pattern of soil moisture changes over the time for drip are detailed.

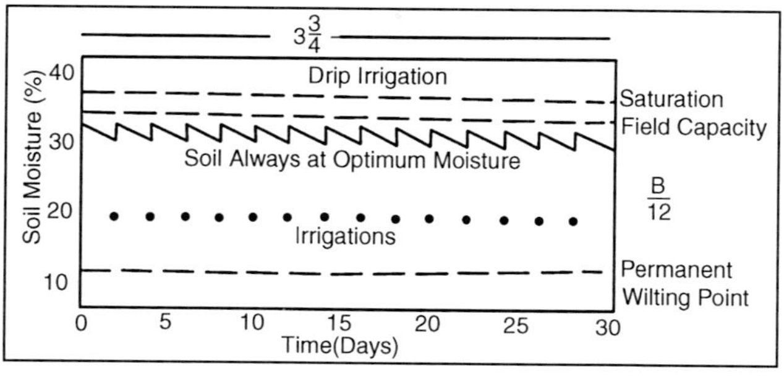

Fig. Drip Irrigation and Soil Moisture Patterns.

ADVANTAGES OF DRIP IRRIGATION

Following are the main advantages of the drip irrigation.

- Controlled application of water as per the needs of plants at low pressure to limited soil areas (root zones).
- Water saving to the tune of 50 to 70 per cent by reducing the total evaporative surface, reduction in runoff and controlling deep percolation losse.
- Soil erosion is minimal, due to no runoff water on surface.
- Weed growth is minimum.
- Water loss through transpiration is low.
- Development of surface crust and determination of surface soil structure is avoided. Soil compaction is less.
- Limited soil wetting permits undisturbed cultural practices.
- It is possible to obtain, better yield and quality of crops by controlling soil moisture-air-nutrients level.
- We can save the fertilizers by monitoring the supply of nutrients as per the need of the crop.
- Improvements in biological fertility can be achieved by avoiding pollution.

ESTABLISHING OF ECONOMICS

The initial expenditure in establishing a long term drip irrigation system usually costs around Rs.32000/- per hectare. This includes the cost of all the components mentioned earlier except the source of water and pump set on the well or boring. This heavy initial expenditure usually discourages the cultivators to install drip system in his fields, though subsidies are given on the purchase of these equipments by the central and state governments. The drip system of irrigation is usually adopted in Maharastra by orchards and sugarcane growers. They receive subsidies from the sugar development funds, besides the other agencies. The comparative figures studied for sub-surface and surface drip and conventional furrow irrigation are given in detail in Table below.

The below data was collected from the trials conducted at Mahatma Phule Agri. University, Rahuri on sugarcane crop. The highest yield (175 tonnes/ha) was obtained in surface drip and paired planting of sugarcane with daily drip irrigation. The yield gain was around 36 per cent over the control yield. The water efficiency use in this treatment was also quite high *i.e.* 1.631. The water saving over the conventional fur- row method was over 50 per cent.

It is calculated that the gain of 47 Tonnes/ha will earn I around ₹ 23,500 per year (@ Rs.500 per tonne). The drip irrigation systems of long term life (five years), may give farmer additional income of around ₹ 1,17,500 which may cover the initial cost. On the long term basis, the drip irrigation system is more economical and paying for the farmers. The drip system could be popularised if growers are financed in the initial stages and subsidies are allowed.

Table: Sugarcane Yield, Quantity of Irrigation Water Applied and its Efficiencies

Treatment No.	Particulars	Total quantity of irrigation water applied (HA-cms)	Yield/ MT/HA	Increase in Yield over control	Water use efficiencies MT/HA cms
T1	Subsurface, paired planting daily irrigation	93.86	143.8	12.22	1.532
T2	Subsurface, paired planting alternate day irrigation	93.86	142.5	11.91	1.518
T3	Subsurface, paired planting daily irrigation	107.56	175.5	36.94	1.631
T4	Subsurface, paired planting alternate day irrigation	187.6	6154.21	20.32	0.821
T5	Furrow method of irrigation, normal planting (control)	215.35	128.16	-	0.595

THE CAPITAL FINANCING OF IRRIGATION INFRASTRUCTURE

The infrastructural supply of irrigation services requires a wide variety of real capital resources. So the actors responsible for service supply must find the capital finance for these resources' capital costs. John Briscoe has pointed out how neglected is the subject of irrigation's capital financing. He writes:

To illustrate just how complete this neglect of financing is, a recent major review of the World Bank experience in irrigation...has this and not a word more, to say about irrigation financing: 'There are no reliable statistics on global irrigation investment'. Similarly, an irrigation policy review for the UK Department for International Development does not mention the issue of financing save for a report on official development assistance flows to the irrigation sector... The two reports to which he refers were published by the World Bank. In this 'data desert' Briscoe himself estimates the aggregate level of irrigation investment in the developing countries. Using a variety of sources, including country-level data published in 1993-6 from Bangladesh, Colombia,

Egypt, Ethiopia, India, Laos, Mexico, Morocco, Nigeria and Vietnam, he arrives at a figure of US$25 billion per year. For water-related infrastructure as a whole, including irrigation, hydropower, water supply and sanitation, the mix of funding is 90 per cent of investment from domestic sources and 10 per cent from external sources.

The World Commission on Dams suggests that the annual investment in large dams for irrigation in the 1990s, excluding their distribution infrastructure, was US$8-11 billion in the developing countries. In the developed countries, investment in large dams for irrigation, water supply and flood control was US$3-5 billion. The WCD used the term 'large dams' to mean a dam with a height of 15 m or more from the foundation as well as dams 5-15 m in height with a reservoir volume of more than 3 mcm. There are 45,000 such dams around the world.

Historically the actors responsible for service supply have fallen into three groups: farming families financing their own infrastructural supply; agribusinesses in a similar position; and public-sector irrigation and drainage authorities. Broadly speaking the first of these has the smallest capital needs per actor; the third has the largest; and agribusinesses lie in the middle. More recently, non-governmental local entities for collective action, such as water user associations, have paid for machinery and equipment out of their irrigation fee income.

The variety of ways in which capital financing is accessed is great, both within and between these three groups. The same applies to the terms under which finance is made available, such as the maximum loan available, the number of years over which it must be repaid, the rate of interest charged and the security demanded in case of failure to meet the agreed repayments schedule. These terms are best understood in the light of each country's banking and credit system or, in the case of many public-sector irrigation and drainage agencies, with reference to the policies and practices of the international banks. Large agribusinesses may access finance not merely through credit but also through capital raised by issuing shares. Briscoe stresses how important is the supply of irrigation water from the private sector. He suggests that it is as high as 70 per cent in Pakistan and 40 per cent in the Philippines and India. He also associates the form of supply appropriation with the institutional source. 'Surface irrigation is the classic government-driven public works and rent-seeking operation... Groundwater irrigation is usually "out of sight" in the informal, small-scale private sector'.

David Sutherland and Peter Howsam have prepared a reference manual on groundwater irrigation technology management that can be used to illustrate investment financing in the case of farming families in Bangladesh and Pakistan. The sources of money for the purchase of pumps and their power units are indeed diverse and include: the farming family's own savings, money borrowed from relatives and friends, the hiring of irrigation equipment, non-governmental organizations (NGOs) providing loans, credit from private or national agricultural banks and loans or cash subsidies from local and regional government. Better-off farmers who succeed in access to credit may supply their surplus irrigation water to smaller farms; this is particularly the case in Pakistan. However, there are difficulties associated with the supply of credit in Bangladesh and Pakistan. Reluctance on the side of the farming family to borrow may derive from the loan repayments they will have to make prior to harvest time, the complexity of loan procedures, the bribes required by bank staff to issue a loan, the indebtedness risks created by fluctuations in crop turnover and, more generally, the stark poverty that makes repayment of any loan prohibitive amongst the landless and the smallest landowners. On the side of lenders, unwillingness to lend is caused by bad debts accruing on their profit and loss account exacerbated by government waiver of farmer debts, borrowing requirements that exceed microcredit limits and the transaction costs associated with large numbers of small loans.

The inability of farming families to mobilize capital restricts the number of entrants to private tubewell ownership or drives many to buy the cheapest technology available, even when the lower capital costs are more than offset by higher costs of operation and maintenance. The WCD suggests that the first and single largest financier of large dams has often been the World Bank. With respect to the developing countries we can expand on this. Briscoe believes that 10 per cent of the financing of water-related infrastructure in those countries comes from external sources. Of this about one-half is provided by the Bank alone. In 1997 the Bank's portfolio of irrigation and drainage projects had commitments valued at about US$7 billion, about 6 per cent of its total portfolio. These commitments are to the third category of irrigation suppliers referred to above-the public-sector irrigation and drainage authorities.

The World Bank's capacity to lend on this prodigious scale exists only because it also has an ability to borrow of extraordinary magnitude. Most of the Bank's lending is long term; so too is the bulk of its borrowing. This takes place in the world market for bonds. In this 8-month period alone the face value of sterling-denominated bonds and US dollar-denominated bonds issued by the Bank was £765 million and US$4,050 million respectively. Of fifteen issues, only five had a maturity date less than 5 years after issue.

The costs borne by the Bank of the bonds it issues are the fees it pays at the time of their marketing and the rate of interest the Bank has to pay on the money it borrows. It is a record for the London market for the last Wednesday in each month (or thereabouts) from August 2000 to March 2001 of the yield on World Bank bonds with a redemption date in June 2010. Here it is 7.125 per cent. But the actual rate of interest earned on a bond with a face value of, say, US$100 depends on *how much the lender pays for such a bond*. This payment may be more or less than the face value. The bid yield on a bond-the rate of interest actually received for a given bid price-is equal to the coupon multiplied by 100 and divided by the bid price. Where the bid price happens to be 100, the yield is equal to the coupon. Simple! It follows that when the World Bank raises fresh capital on a bond denominated say in US dollars, with a redemption date say in June 2010, it knows that the actual rate of interest it will have to pay on that new money is going to be about equal to the current bid yield on bonds of that type.

The market rate of interest on bonds with a long-term date has many determinants. Just one is highlighted here and that is the credit rating of the institution issuing the bond. Credit rating is provided by two major agencies: Standard and Poor's and Moody's.

As a result the market is willing to accept a rate of interest from the Bank on newly issued capital denominated in US dollars only 100 or so basis points above that paid by the safest security in the world, bonds issued by the US Treasury. This 'spread' is given in column 8. It ranges from 73 to 116 basis points, *i.e.* from 0.73 per cent to 1.16 per cent. With respect to irrigation (and other) projects, the institution that borrows from the Bank is typically national government, not the institution that is responsible for the construction and management of the new infrastructure. An intermediary stands between the Bank and the irrigation authority. As a result, if the project fails the Bank does not suffer financially. There is a break in the line of prudential responsibility for ensuring that the institution requiring loan finance for capital investment is capable of repaying its debt.

GSTATION PERIOD DELAYS AND COST OVERRUNS

The gestation period of an irrigation project is the length of time from the inception of design and construction to completion. The World Commission on Dams has published useful data and analysis of gestation period delays with large dams and their associated investment cost overruns. Dams in the WCD Knowledge Base demonstrate a marked tendency to both delay and overrun with interesting interdependence of the engineering and financing determinants. Note that there are two main categories of large dam. *First,* the reservoir-type storage dam impounds water behind the dam for seasonal, annual and in some cases multi-annual storage and regulation of the river. *Second,* the run-of-river dam often has no storage reservoir and may have limited daily pondage; it creates a hydraulic head in the river to divert some portion of the river flow to a canal or power station.

Large dam *delays* are considered first. In comparing the planned gestation period with actual outcome of the ninety-nine projects in the WCD's cross-check survey, fifty came in on or before schedule, twenty-seven were delayed for 1 or 2 years, eleven for 3 or 4 years and four for 5 or 6 years. Four dams were delayed for more than 10 years. 'The existing literature on large dams and related projects confirms this finding: large dams tend to be subject to significant schedule slippages'. The causes of delay are cited as contractor and construction management inefficiency, unrealistic construction schedules, late delivery to site of essential equipment, labour unrest, legal challenges by affected groups and, finally, difficulties in raising capital finance for the engineering work, as with the Tucurui Dam in Brazil.

The impact of gestation period delay is to postpone both the delivery of the dam's services as well as the inception of the stream of revenues accruing to the dam owner. Economic and financial performance is consequently weakened. Delay also brings with it an increase in the interest charges on the capital borrowed over the gestation period. In the case of Tucurui, for example, more than one-third of its 77 per cent cost overrun is attributable to additional interest payments during construction owing to construction delay. The general issue of *cost overruns* is now considered. These are serious. The variability of outcome is high, particularly with multi-purpose projects. There is no clear correlation with dam size. Dams built in India perform worst according to the WCD report. But the overall results are less dramatic when corrected for inflation during the gestation period, falling from an average exceeding 50 per cent at out-turn prices down to 21 per cent in 1998 US dollars. The causes of investment cost overrun include poor development of cost estimates, technical problems arising during construction, weak implementation by contractors and poor prediction of inflation. The WCD comments:

Part of the difficulty in developing accurate projections for construction costs of large dams is that the geotechnical conditions at a site (the quality of the rock for the foundations of the major structure and for tunnels) and the quality of the construction materials cannot be determined precisely until construction is under way. Discovery during construction of less favourable site conditions than those assumed in the engineering designs and construction plans can be a significant contributor to cost overruns and delays in time schedules. Despite being a common factor in causing overruns, little to no provisions have been made to improve the estimates in this regard.

The WCD suggests its evidence confirms that there is a systematic bias towards underestimation of large dams' capital costs. Dam projects are usually given the go-ahead on the basis of a financial budget for the investment. The impact of substantial overruns is that additional funds have to be found, sometimes creating difficulties for public or private budgeting. Where tariffs are based on cost estimates, overruns can undermine cost recovery and the project's financial rate of return.

Case Study: European Investment Bank Activity in the Mediterranean

This case study of project financing by the European Investment Bank (EIB) in the Mediterranean takes up the issue of prudential responsibility. The

EIB was set up in 1958 by the Treaty of Rome, is owned by the fifteen European member states and has its headquarters in Luxembourg. It supports European Union (EU) policies on a self-financing basis, raising its resources on the world's capital markets for onlending, it says, to sound capital investment projects that promote the balanced development of the EU. As a major international borrower that has always been awarded the highest AAA credit rating by the world's leading rating agencies, the Bank raises large volumes of funds on excellent terms. It onlends the proceeds of its borrowings on a non-profit basis. While the bulk of its loans are within the European Union, the Bank has also been called upon to participate in the implementation of the Union's development aid and co-operation policies in some 120 non-EU countries.

The EIB's financing activities in the EU's member countries in the Mediterranean began with the Bank's inception. Financing in non-member Mediterranean countries began in 1962 and has chiefly taken the form of long-term loans, although money has also been made available as risk capital funding designed to encourage development of the local private sector and in joint ventures. The countries forming the boundary of the Mediterranean Sea and its island states share a broadly similar climate. The summers are dry and hot and the rainfall pattern can result either in severe floods or in droughts.

Precipitation varies from an annual average of 1,000 mm in the mountainous areas of the north to less than 100 mm in the south. A wide diversity in the per capita water resource exists; it is generous in some countries of the Balkan sub-region but arid in countries such as Malta and Tunisia. Water use per capita also varies considerably, from 100 m^3/ year at the low end of the range to more than 1,000 m^3/year in countries with a large irrigated sector. It is the southern and eastern countries in the region that most experience water stress.

This has led to the mining of deep aquifers, sea water desalination and the use of land-sourced brackish water at high unit costs. 'The situation is particularly serious where the environment is under severe threat, where expensive water is wasted through distribution pipe leakages, rationed by intermittently cutting the supply to consumers, or wasted through inefficient irrigation systems and practices'. The Bank believes that the reuse of treated waste water for irrigation and industrial purposes has an important part to play in the development of integrated solutions for water resource and environmental pollution problems.

In 1999 the EIB published a review by its Evaluation Unit and by outside consultants of seventeen water projects it had financed around the Mediterranean Sea mainly during the 1980s. This case study is based on that report. The projects selected are located in eight countries north and south of the Mediterranean, they accounted for 11 per cent of the projects financed by the Bank from 1981 to 1992 and they represented a total investment cost of 1.4 billion European Currency Units (ECU) and total EIB loans of 430 million ECU. Forty per cent of the investment went for five irrigation projects and the remainder to water resources development, water distribution, sewerage and wastewater treatment. Projects selected were drawn from both EU member countries and those outside the Union. Water projects within the EU are usually financed with the objective of fostering regional development and/or environmental protection. Projects outside the EU are financed under mandates given to the EIB by the European Council in order to promote economic development.

The investment total of 1.4 billion ECU included land acquisition, engineering and supervision, civil works, supply of equipment, working capital and interest during construction. The irrigation projects included extension services and these projects' cultivated areas ranged from a minimum of 200 ha to a maximum of 65,000 ha with a median value of 6,000 ha. Schul writes: Three out of the 5 irrigation projects were implemented according to the original design capacity. One particularly risky project seeking to introduce drip irrigation to stop over-exploiting a non-renewable water source failed entirely, despite intensive EIB technical assistance. In the end, agricultural rather than irrigation equipment was purchased to use up the EIB loan. The technology proposed was over-sophisticated for the local management and the EIB became aware of corruption after the loan was fully disbursed. The institutional set-up has since been changed. Of the seventeen projects, only three were completed on time; the implementation delay on the others ranged from 2 to 12 years. The problem was particularly acute in two EU countries. Schul comments:

Unreasonably slow implementation resulting from institutional weaknesses is one of the main curses of water projects in the Mediterranean area. This has been one of the factors contributing to the EIB's decision to no longer finance irrigation projects in some countries. In one symptomatic case, for instance, the EIB participated with a local bank in an innovative deal where farmers agreed to finance 65 per cent of the investment costs of collective irrigation schemes.

Farmers expected this to solve the problem that 100 per cent subsidised schemes were being continually postponed because of their Government's budgetary restrictions. However, after a few years they withdrew from the arrangement, partly because the Government interfered with procurement procedures, which resulted in delaying access to irrigation water by sometimes several cropping seasons.

In the meantime the Government managed to accelerate its own programme with EU grants providing 100 per cent subsidies. Six of the seventeen projects experienced difficulties common to public procurement procedures.

These were:

- Excessive delays between bid receipt and contract award,
- Bids accepted from inexperienced contractors or extremely low-cost tenders forcing subsequent renegotiations, quality problems and implementation delays,
- Changes in design, disputes with suppliers and the bankruptcy of appointed contractors.

Investment costs per hectare of irrigation works co-financed with the ultimate beneficiaries were 50 per cent or less of scheme costs fully financed by state authorities. In four of the five irrigation projects, production was equal to or greater than that originally forecast. The fifth project produced nothing. Two projects were limited to smaller irrigated areas than first planned because rapid urbanization reduced both the available supply of water and the farm hectarage. Three of the five projects made possible a reduction in desertification or soil salinization or soil erosion on steep slopes. Treated waste water was reused in two of the projects for irrigation supply.

The loan characteristics of the seventeen projects deserve consideration and they are set out in Table. Project investment cost, the rate of interest, the grace period and the repayment period are variable to a staggering degree. The huge gap between the minimum and median values of the loan rate of interest deserves explanation. The EIB's own policy on external lending is to charge a standard mark-up: 25 basis points above the cost of borrowing. But the EU budget provides grant funds for a number of the mandates, including the Mediterranean, part of which is available to subsidize the interest rate on EIB loans. The EIB states that it lends money on 'sound capital investment projects'. But the minimum rates of return in 1981-92 was willing to lend on a group of projects that included cases of both negative financial and negative economic returns in *ex ante* terms. What is equally startling is that the median *ex post* financial return is negative. Finally the median *ex post* economic return is a full 320 basis points below the *ex ante* calculations. Low financial rates of return are attributable variously to the cumbersome decision-making procedures of the public sector, excessive operating costs, low charges for water, the disbursing of loans at loan signature and excessive implementation delays. In only two out of the seventeen projects did financial profitability match EIB lending rates.

The Review's summary points out that most water resource projects and agencies are financially unprofitable, with irrigation projects not performing significantly worse than the other sub-sectors. In respect of irrigation projects within the EU farmers deviated from the original cropping patterns and adopted crops subsidized by the Common Agricultural Policy. Here farmers pursued financially rewarding activities, but ones that, from the point of view of the Union as a whole, had negative economic returns. We have noted that the EIB is a self-financing institution.

How then is it feasible for the Bank:

- To raise money in the international bond market on strictly commercial terms;
- To on-lend these same funds at interest rates that cover both the Bank's own debt repayment obligations as well as its costs of operations;
- Over an 11-year period to appraise and approve loans on seventeen water resource projects that in retrospect enjoyed financial profitability that matched the Bank's lending rates in only two cases and where the median *ex post* financial rate of return was negative?

To answer this question we must return to the issue of prudential responsibility and its fragmentation. The institutional category of both the borrowers of the EIB's loans and the projects' promoters. Apparently all these institutions were in the public sector, even the Société Anonyme. Only two loans were granted directly to the project promoter; the remainder went first to an intermediary that channelled funds on to the promoter. In thirteen cases the borrower was a central government ministry or a state bank.

Typically, the full loan is disbursed at loan signature or in tranches with a view to matching expenditure. This is particularly the case within the European Union, where the EIB's loans are made for projects in an advanced state of implementation. The Bank's records show that, after the initial project appraisal, EIB staff visited water projects in the Mediterranean *only once in every 5.5 years*. Loan provision to government or some other intermediary usually means that the financial terms on which the loan is made to the middle-man is quite different from the terms between that middle-man and the beneficiary of the loan.

The fragmented relationship between the Bank and the final recipient of a loan exists, then, because loans are typically made through an intermediary, the loan is disbursed on signature and EIB follow-up is weak. But there is a fourth, complex issue that arises from the channelling of Bank loans through government (or any other) intermediary. This is known as the 'fungibility' issue and it requires explanation. Imagine that a multilateral finance institution (MFI) such as the EIB or the World Bank lends money to an irrigation authority for construction of new storage requirements. The MFI stains the notes red and delivers them directly to the authority (there is no intermediary) and insists in the financial agreement (which is honoured) that all these red notes and only these red notes are spent on the new infrastructure. In this unusual case we have no doubt whatsoever that the loan does indeed finance the project.

But the world does not spin like this, as we have seen. What actually happens is that the MFIs loan goes to an intermediary such as a government department that probably places the money in an interest-earning account where it joins financial inflows from a multiplicity of other sources. When the irrigation authority draws down the loan made to it for dam construction, there is no direct link between the financial transfer from department to irrigation authority and the original loan from the MFI to the department. Indeed it may well be the case that the department-to-irrigation authority loan would have gone ahead even in the absence of the MFI-to-department loan. In such cases we speak of the MFI loan as meeting the 'treasury requirements' of the department. The non-existence of a direct link between the money loaned by the MFI and the money borrowed by the irrigation authority is an instance of what is called 'fungibility'.

Here is a hydrosocial parallel. Three streams feed a reservoir. The waters from the dam are channelled along a river past seven farms that divert the flow for crop irrigation. There is no sense in saying that any single stream sources any single farm. The *Concise Oxford Dictionary* says of the adjective 'fungible' that it is a term in law, with reference to goods contracted for, when an individual specimen is not meant. In this context 'fungible' means 'that can serve for, or be replaced by, another answering to the same definition'.

The central conundrum of fungibility is as follows. However tightly an MFI ties its financing to a specific project, a national government will have a wide range of projects for which it is seeking finance. If the MFI provides capital finance for a high-priority project that government would have managed to finance anyway, then the MFI loan permits government to transfer the funding released to the highest priority for which funds are *not* available, one which the MFI may or may not approve of. However, if the MFI provides capital finance for a low-rated project, truly additional funding in this case, that project may well not be one which the MFI wishes to see go ahead. As Tony Faint writes (personal communication), this is a dilemma for all project-based lending and the EIB is essentially a project-financing institution.

Schul's excellent report makes all of this clear. He recognizes the uncertainty in the meaning of the Bank's 'contribution to objectives through its investments' when projects would have gone ahead even in the absence of Bank financing. Whilst the Bank essentially makes loans to sound, creditworthy borrowers, this may have contributed to weakening the link between EIB loans and the projects.

Loans were usually channelled through intermediaries and resources were then allocated to the final beneficiary on the intermediary's own terms, rather than on the EIB's. Hence, lending terms were either not well adapted to the investment or the promoter, or EIB funds *de facto* financed the treasury requirements of the intermediary institution or public administration. This may have reduced the EIB's financial exposure, but its scope to influence the outcome of projects and impose the discipline that usually accompanies bank lending was also diminished. The Report's conclusions include the view that the link between its loans and the targeted investments needs to be tightened in order for the Bank to be able to apply appropriate loan conditions and bank discipline on promoters. Loans should be channelled as much as possible to the promoter. In addition, projects should be monitored at least once per year with regular reports to management on specific portfolios.

A RELIABLE AND SUITABLE IRRIGATION WATER SUPPLY

A reliable and suitable irrigation water supply can result in vast improvements in agricultural production and assure the economic vitality of the region. Many civilizations have been dependent on irrigated agriculture to provide the basis of their society and enhance the security of their people. Some have estimated that as little as 15-20 per cent of the worldwide total cultivated area is irrigated. Judging from irrigated and non-irrigated yields in some areas, this relatively small fraction of agriculture may be contributing as much as 30-40 per cent of gross agricultural output.

Effective agronomic practices are essential components of irrigated systems. Management of the soil fertility, cropping selection and rotation, and pest control may make as much incremental difference in yield as the irrigation water itself. Irrigation implies drainage, soil reclamation, and erosion control. When any of these factors are ignored through either a lack of understanding or planning, agricultural productivity will decline. History is absolutely certain on this point.

Irrigated agriculture faces a number of difficult problems in the future. One of the major concerns is the generally poor efficiency with which water resources have been used for irrigation. A relatively safe estimate is that 40 per cent or more of the water diverted for irrigation is wasted at the farm level through either deep percolation or surface runoff. These losses may not be lost when one views water use in the regional context, since return flows become part of the usable resource elsewhere. However, these losses often represent foregone opportunities for water because they delay the arrival of water at downstream diversions and because they almost universally produce poorer quality water. One of the more evident problems in the future is the growth of alternative demands for water such as urban and industrial needs. These uses place a higher value on water resources and therefore tend to focus attention on wasteful practices. Irrigation science in the future will undoubtedly face the problem of maximizing efficiency.

Irrigation in arid areas of the world provides two essential agricultural requirements: (1) a moisture supply for plant growth which also transports essential nutrients; and (2) a flow of water to leach or dilute salts in the soil. Irrigation also benefits croplands through cooling the soil and the atmosphere to create a more favourable environment for plant growth.

The method, frequency and duration of irrigations have significant effects on crop yield and farm productivity. For example, annual crops may not germinate when the surface is inundated causing a crust to form over the seed bed. After emergence, inadequate soil moisture can often reduce yields, particularly if the stress occurs during critical periods. Even though the most important objective of irrigation is to maintain the soil moisture reservoir, how this is accomplished is an important consideration. The technology of irrigation is more complex than many appreciate. It is important that the scope of irrigation science not be limited to diversion and conveyance systems, nor solely to the irrigated field, nor only to the drainage pathways. Irrigation is a system extending across many technical and non-technical disciplines. It only works efficiently and continually when all the components are integrated smoothly.

IRRIGATION METHODS AND THEIR SELECTION

There are three broad classes of irrigation systems: (1) pressurized distribution; (2) gravity flow distribution; and (3) drainage flow distribution. The pressurized systems include sprinkler, trickle, and the array of similar systems in which water is conveyed to and distributed over the farmland through pressurized pipe networks.

There are many individual system configurations identified by unique features (centre-pivot sprinkler systems). Gravity flow systems convey and distribute water at the field level by a free surface, overland flow regime. These surface irrigation methods are also subdivided according to configuration and operational characteristics. Irrigation by control of the drainage system, subirrigation, is not common but is interesting conceptually. Relatively large volumes of applied irrigation water percolate through the root zone and become a drainage or groundwater flow. By controlling the flow at critical points, it is possible to raise the level of the groundwater to within reach of the crop roots. These individual irrigation systems have a variety of advantages and particular applications which are beyond the scope of this paper. Suffice it to say that one should be familiar with each in order to satisfy best the needs of irrigation projects likely to be of interest during their formulation.

Irrigation systems are often designed to maximize efficiencies and minimize labour and capital requirements. The most effective management practices are dependent on the type of irrigation system and its design. For example, management can be influenced by the use of automation, the control of or the capture and reuse of runoff, field soil and topographical variations and the existence and location of flow measurement and water control structures. Questions that are common to all irrigation systems are when to irrigate, how much to apply, and can the efficiency be improved. A large number of considerations must be taken into account in the selection of an irrigation system. These will vary from location to location, crop to crop, year to year, and farmer to farmer. In general these considerations will include the compatibility of the system with other farm operations, economic feasibility, topographic and soil properties, crop characteristics, and social constraints.

Compatibility

The irrigation system for a field or a farm must function alongside other farm operations such as land preparation, cultivation, and harvesting. The use of the large mechanized equipment requires longer and wider fields. The irrigation systems must not interfere with these operations and may need to be portable or function primarily outside the crop boundaries (*i.e.* surface irrigation systems). Smaller equipment or animal-powered cultivating equipment is more suitable for small fields and more permanent irrigation facilities.

Economics

The type of irrigation system selected is an important economic decision. Some types of pressurized systems have high capital and operating costs but may utilize minimal labour and conserve water. Their use tends toward high value cropping patterns. Other systems are relatively less expensive to construct and operate but have high labour requirements. Some systems are limited by the type of soil or the topography found on a field. The costs of maintenance and expected life of the rehabilitation along with an array of annual costs like energy, water, depreciation, land preparation, maintenance, labour and taxes should be included in the selection of an irrigation system.

Topographical characteristics

Topography is a major factor affecting irrigation, particularly surface irrigation. Of general concern are the location and elevation of the water supply relative to the field boundaries, the area and configuration of the fields and access by roads, utility lines (gas, electricity, water, etc.), and migrating herds whether wild or domestic. Field slope and its uniformity are two of the most important topographical factors. Surface systems, for instance, require uniform grades in the 0-5 per cent range.

Soils

The soil's moisture-holding capacity, intake rate and depth are the principal criteria affecting the type of system selected. Sandy soils typically have high intake rates and low soil moisture storage capacities and may require an entirely different irrigation strategy than the deep clay soil with low infiltration rates but high moisture-storage capacities. Sandy soil require more frequent, smaller applications of water whereas clay soils can be irrigated less frequently and to a larger depth. Other important soil properties influence the type of irrigation system to use. The physical, biological and chemical interactions of soil and water influence the hydraulic characteristics and filth. The mix of silt in a soil influences crusting and erodibility and should be considered in each design. The soil influences crusting and erodibility an should be considered in each design. The distribution of soils may vary widely over a field and may be an important limitation on some methods of applying irrigation water.

Water supply

The quality and quantity of the source of water can have a significant impact on the irrigation practices. Crop water demands are continuous during the growing season. The soil moisture reservoir transforms this continuous demand into a periodic one which the irrigation system can service. A water supply with a relatively small discharge is best utilized in an irrigation system which incorporates frequent applications. The depths applied per irrigation would tend to be smaller under these systems than under systems having a large discharge which is available less frequently. The quality of water affects decisions similarly. Salinity is generally the most significant problem but other elements like boron or selenium can be important. A poor quality water supply must be utilized more frequently and in larger amounts than one of good quality.

Crops

The yields of many crops may be as much affected by how water is applied as the quantity delivered. Irrigation systems create different environmental conditions such as humidity, temperature, and soil aeration. They affect the plant differently by wetting different parts of the plant thereby introducing various undesirable consequences like leaf burn, fruit spotting and deformation, crown rot, etc. Rice, on the other hand, thrives under ponded conditions. Some crops have high economic value and allow the application of more capital-intensive practices. Deep-rooted crops are more amenable to low-frequency, high-application rate systems than shallow-rooted crops.

Social Influences

Beyond the confines of the individual field, irrigation is a community enterprise. Individuals, groups of individuals, and often the state must join together to construct, operate and maintain the irrigation system as a whole. Within a typical irrigation system there are three levels of community organization. There is the individual or small informal group of individuals participating in the system at the field and tertiary level of conveyance and distribution.

There are the farmer collectives which form in structures as simple as informal organizations or as complex as irrigation districts. These assume, in addition to operation and maintenance, responsibility for allocation and conflict resolution. And then there is the state organization responsible for the water distribution and use at the project level. Irrigation system designers should be aware that perhaps the most important goal of the irrigation community at all levels is the assurance of equity among its members. Thus the operation, if not always the structure, of the irrigation system will tend to mirror the community view of sharing and allocation. Irrigation often means a technological intervention in the agricultural system even if irrigation has been practiced locally for generations. New technologies mean new operation and maintenance practices. If the community is not sufficiently adaptable to change, some irrigation systems will not succeed.

External Influences

Conditions outside the sphere of agriculture affect and even dictate the type of system selected. For example, national policies regarding foreign exchange, strengthening specific sectors of the local economy, or sufficiency in particular industries may lead to specific irrigation systems being utilized. Key components in the manufacture or importation of system elements may not be available or cannot be efficiently serviced. Since many irrigation projects are financed by outside donors and lenders, specific system configurations may be precluded because of international policies and attitudes.

ADVANTAGES AND DISADVANTAGES OF SURFACE IRRIGATION

The term 'surface irrigation' refers to a broad class of irrigation methods in which water is distributed over the field by overland flow. A flow is introduced at one edge of the field and covers the field gradually. The rate of coverage (advance) is dependent almost entirely on the differences between the discharge onto the field and the accumulating infiltration into the soil. Secondary factors include field slope, surface roughness, and the geometry or shape of the flow cross-section. The practice of surface irrigation is thousands of years old. It collectively represents perhaps as much as 95 per cent of common irrigation activity today. The first water supplies were developed from stream or river flows onto the adjacent flood plain through simple check-dams and a canal to distribute water to various locations where farmers could then allocate a portion of the flow to their fields. The low-lying soils served by these diversions were typically high in clay and silt content and tended to be most fertile. The land slope was normally small because of the structure of the flood plain itself. With the advent of modern equipment for moving earth and pumping water, surface irrigation systems were extended to upland areas and lands quite separate from the flood plain of local rivers and streams. These lands tend to have more variable soils and topographies, are usually better drained, and may be naturally less fertile. Thus, these lands usually require greater attention to design and operation.

Advantages

Surface irrigation offers a number of important advantages at both the farm and project level. Because it is so widely utilized, local irrigators generally have at least minimal understanding of how to operate and maintain the system. In addition, surface systems are often more acceptable to agriculturalists who appreciate the effects of water shortage on crop yields since it appears easier to apply the depths required to refill the root zone.

The second advantage of surface irrigation is that these systems can be developed at the farm level with minimal capital investment. The control and regulation structures are simple, durable and easily constructed with inexpensive and readily-available materials like wood, concrete, brick and mortar, etc. Further, the essential structural elements are located at the edges of the fields which facilitates operation and maintenance activities. The major capital expense of the surface system is generally associated with land grading, but if the topography is not too undulating, these costs are not great. Recent developments in surface irrigation technology have largely overcome the irrigation efficiency advantage of sprinkler and trickle systems. An array of automating devices roughly equates labour requirements. The major trade-off between surface and pressurized methods lies in the relative costs of land levelling for effective gravity distribution and energy for pressurization. Energy requirements for surface irrigation systems come from gravity. This is a significant advantage in today's economy.

Another advantage of surface systems is that they are less affected by climatic and water quality characteristics. Even moderate winds can seriously reduce the effectiveness of sprinkler systems. Sediments and other debris reduce the effectiveness of trickle systems but may actually aid the performance of the surface systems. Salinity is less of a problem under surface irrigation than either of these pressurized systems. There are other advantages specific to individual regions that might be mentioned. Surface systems are better able to utilize water supplies that are available less frequently, more uncertain, and more variable in rate and duration. The gravity flow system is a highly flexible, relatively easily-managed method of irrigation.

Disadvantages

There is one disadvantage of surface irrigation that confronts every designer and irrigator. The soil which must be used to convey the water over the field has properties that are highly varied both spatially and temporally. They become almost undefinable except immediately preceding the watering or during it. This creates an engineering problem in which at least two of the primary design variables, discharge and time of application, must be estimated not only at the field layout stage but also judged by the irrigator prior to the initiation of every surface irrigation event. Thus while it is possible for the new generation of surface irrigation methods to be attractive alternatives to sprinkler and trickle systems, their associated design and management practices are much more difficult to define and implement.

Although they need not be, surface irrigation systems are typically less efficient in applying water than either sprinkler or trickle systems. Many are situated on lower lands with heavier soils and, therefore, tend to be more affected by waterlogging and soil salinity if adequate drainage is not provided. The need to use the field surface as a conveyance and distribution facility requires that fields be well graded if possible. Land levelling costs can be high so the surface irrigation practice tends to be limited to land already having small, even slopes.

Surface systems tend to be labour-intensive. This labour need not be overly skilled. But to achieve high efficiencies the irrigation practices imposed by the irrigator must be carefully implemented. The progress of the water over the field must be monitored in larger fields and good judgement is required to terminate the inflow at the appropriate time. A consequence of poor judgement or design is poor efficiency.

One sometimes important disadvantage of surface irrigation methods is the difficulty in applying light, frequent irrigations early and late in the growing season of several crops. For example, in heavy calcareous soils where crust formation after the first irrigation and prior to the germination of crops, a light irrigation to soften the crust would improve yields substantially. Under surface irrigation systems this may be unfeasible or impractical as either the supply to the field is not readily available or the minimum depths applied would be too great.

5

Uses and Impacts of Irrigation Water

Irrigation impacts may vary considerably depending on the source of water, river diversion project or from large reservoir or groundwater based irrigation. Groundwater is mainly used for local level crop production and municipal uses.

However, surface irrigation infrastructure provides water for a variety of uses with wider area coverage, with potential impacts and likelihood of associated externalities greater than that from groundwater use. Water uses from surface irrigation water can be classified as follows:

Withdrawal uses:

- Irrigated agriculture
- Urban domestic uses
- Commercial and industrial uses
- Rural domestic and livestock, and rural small-scale enterprises.

In-stream uses:

- Commercial/recreational fish production
- Hydropower generation
- Water transportation
- Recreational uses of river flows, swimming, river rafting, etc.
- Ecological uses, aquatic ecosystem protection, landscape view, etc.
- Religious use of river, bathing, etc.

IMPACTS/OUTCOMES OF IRRIGATION WATER USES

In most of the above uses, irrigation water has multiple effects. Some of the irrigation induced impacts are desirable and beneficial to society while others are undesired and adverse in terms of their negative impacts on humans and environment (ecosystems). Some of the impacts are significant, while others may be insignificant—some of them are known while others remain unknown. Some of the impacts are common to most irrigation development projects, while others are more specific to certain locations, schemes or methods of irrigation. These impacts may be broadly classified into desired impacts (benefits) and undesired or negative impacts (costs). In a project level comprehensive assessment of irrigation impacts, following major impacts, both positive and negative impacts, can be identified.

Major Positive Impacts

(I) direct positive impacts of irrigation include:

- Increased agricultural production
  - Increased crop productivity
  - Expansion in crop areas
  - Increase in cropping intensity
  - Increase in crop diversification
- Increased commercial fish production (in-land fisheries)
- Increased benefits of water use in industrial, commercial and residential sectors-from raw water provided through irrigation infrastructure or from groundwater
- Increased environmental benefits of water for in-stream flows, disposal of waste, wildlife, flora and fauna; increased farm forestry and vegetation in irrigated areas.
- Increased health benefits-improved sanitation due to better access to water.
- Other direct positive impacts
  - Increased benefits from flood control
  - Increased benefits from water use for rural domestic and livestock purposes
  - Increased groundwater recharge; reduction in opportunity costs of water uses
  - Increased recreation from water bodies, sight seeing, fishing

Secondary Impacts of Irrigation

- Increased employment in agriculture due to increased cropping intensity, increased crop area and output from irrigation
- Increased employment outside agriculture from increased crop output in related industries such as input industry (backward linkages) and output processing industries (forward linkages)
- Positive impact on poverty reduction through increased productivity and increased employment opportunities
- Increased food security at national, regional and local levels
- Lower food prices for consumers, due to productivity gains and increased overall food supplies
- Improved nutrition, improved calorie intake and improved health

Major Negative Impacts

Irrigation may have several potential negative impacts on agro-,eco- and human systems.

These impacts may be also broadly divided into three categories:

1. *Adverse economic impacts:* Higher subsidy, distorted market and a relative neglect of rainfed farming and other less favored farming practices
2. *Adverse social impacts:* Forced displacement and involuntarily resettlement of the local inhabitants
3. *Adverse environmental impacts:* Loss of biodiversity, obstruction on natural hydrological flows and damages to aquatic ecosystems

Some of the impacts are direct and enter into normal market process, whereas other negative impacts may be indirect and their costs may not be accounted for by individual decision-makers, nor are these costs are captured into the normal marketing activities— they are called as "externality effects." The term "externality" is generally defined as an effect or outcome when production or consumption of one party affects the production or consumption of another party and neither party makes any compensation for the effect. For example, costs imposed on downstream users of irrigated water polluted by upstream users. These are costs imposed on a society as a result of an action—irrigation development in this case. Based on the nature of the impacts on the surrounding environment and the changes they will bring in the hydrology and in the surrounding regions, environmental impacts or externalities generated by irrigation can be broadly grouped into three categories.

They are:

- *Change in water quantity:* This is related to where water goes and how it is used, *e.g.*, localized scarcity, water stress, etc.
- *Change in water quality:* This is related to change in quality and usability of water, *e.g.*, water contamination, salt loading, etc.
- *Change in soil quality:* This is related to impacts of water uses on land fertility and changes in land quality, *e.g.*, salinity build up, impacts on soil quality by changes in the water table, etc.

Some of the potential negative externalities associated with irrigation may include the following:

- Irrigation-induced land degradation
  - Soil salinity and water logging-on-farm and off-farm impacts
  - Loss of soil fertility due to irrigation induced crop intensification
  - Increase in biological imbalances due to irrigation (weeds, pests)
- Surface water pollution-nutrients/chemicals
- Groundwater pollution-nutrients/chemicals
- Toxic concentration of substances in surface and groundwater-salts, metals and pesticides
- Saline return flows
- Health impacts in terms of increased water borne diseases (schistosomiasis, malaria)

- Loss of bio-diversity-birds, fish and other wildlife species extinction
- Negative impacts on wetlands
- Social impacts
  - Communities displaced by large scale irrigation development
  - Loss of cultural heritage and historical places
  - Displacement of unskilled labour in mechanized irrigation

Given the wide range of these impacts, it may be useful to identify the causal factors for these impacts *i.e.* whether the impacts are due to farm level agricultural practices, field level water application systems, water distribution systems, water supply or drainage systems or large reservoirs or tanks.

THE CHALLENGE OF IMPROVING IRRIGATION PERFORMANCE MANAGEMENT

Much has been done in turnover, ISF and other programmes to develop formal WUA, as a means and embodiment of participation in irrigation management. It has been hoped that once established, WUA would implement their roles in irrigation O&M as expected by government, and so contribute to better irrigation performance. Experience suggests that formal WUA have often not achieved what has been expected.

Formalization of WUA has been overemphasized and inadequate attention has been paid to the ways in which PIM has actually helped improve performance. The following paragraphs outline some of the problems in the sustainability of formal WUA and the need for better information about performance impacts of PIM. The most important challenge for development of participatory irrigation management lies in instituting better approaches to perform the key processes of equitably distributing water and maintaining irrigation systems.

FORMAL WUA DISAPPEAR

Registration of formal WUA has involved a great amount of work in terms of local meetings and processing of the documents to have WUA constitutions and by-laws registered with district offices. However it appears that in many, if not most, cases, once a project is completed, local irrigation management quickly reverts to the patterns which had been used before. This may be informal cooperation among neighbouring farmers, dealing with problems on an ad hoc basis.

Often village government plays the major role in selecting the people who actually distribute water and maintain canals, as well as in providing the authority needed to enforce rules, collect money and mobilize labour. Formal WUA, developed as instruments of government projects, often quickly disappear. By necessity, irrigated farmers continue to be actively involved in irrigation management, cleaning canals, distributing water and carrying out other tasks, through various formal and informal institutions.

Project Experience

WUAs in the Madiun irrigation system in East Java had facilitated participation in physical improvements, but one year afterwards WUA were no longer active and irrigation management was an individual or small group matter. WUAs established by the government in the High Performance Sederhana Irrigation Systems (HPSIS) project did not survive. Irrigation was managed using previous organizations except in one site where there had been no previous irrigation. Factors contributing to the disappearance of formal WUA included the lack of mechanisms to assist WUA once community organizers left the system and the emphasis on organizing WUA to carry out government instructions. Similar patterns of formal water user organizations not being sustained have been prevalent in other countries.

Rethinking Sustainability

There is a need for rethinking current approaches to WUA development. While the structures and canals of irrigation systems may appear permanent, actually they are very subject to erosion, flood and other forces bringing change. The sustainability of irrigation systems ultimately lies in the accomplishment of the key tasks needed to keep the systems functioning. Irrigation development requires maintaining and improving management institutions, software, not just adding hardware. Even without changing the current policy framework, it would be possible to develop a much more efficient and effective approach to sustaining participation in irrigation management.

This could focus more directly on how to improve and sustain participation in the key tasks of equitably distributing water and maintaining irrigation infrastructure. As a key element of better focused approaches, annual joint walkthroughs to review O&M and discuss ways to improve performance appear to be essential if participation in irrigation management is to be sustained and improved, both in ISF areas and turned-over systems.

Participation and Performance Management

Participation should be not just something done for its own sake, but a means for improving the performance of irrigation. There is a shortage of information about what difference participatory irrigation management has actually made in irrigation performance, crop yields and the lives of irrigated farmers. Programmes have had a technical focus on physical works. Except for the Starter Project, participatory programmes have paid little attention to optimizing and assessing performance in terms of agricultural impacts.

Measuring Performance

Monitoring of turnover implementation has focused on physical and financial progress of construction and fulfilment of formal requirements for

WUA registration. Monitoring of ISF has concentrated on collection rates as the key indicator. By contrast, there has been much less attention to monitoring key steps for WUA consultation in turnover and ISF.

Aside from a few studies, little systematic information is available on performance impacts in terms of equity in water delivery and changes in crop yields. Field visits do indicate improvements in block level water distribution, but these are not shown well in current monitoring and evaluation systems. Routine monitoring is supposed to categorize WUA as developed, developing or inactive. This emphasizes compliance with formal procedures for meetings, planning and record keeping, with little information on more direct performance indicators.

In general, monitoring is not strongly focused on showing changes in irrigation performance. Equity of water distribution within areas of WUA responsibility is not measured in current WUA monitoring systems. Rotation between different units is usually the main approach applied during periods of shortage.

Formal WUA have fallen far short of government expectations in terms of collection of routine fees for internal WUA use. Collection of routine fees by WUA differs from the episodic approach traditionally used to collect funds for specific works, usually at the start of the season, and pay those who distributed water, once crops have been harvested. Monitoring information is not collected on the resources mobilized by WUA, or the amount of work done to repair or improve physical infrastructure. Information on crop yields is not collected as part of WUA monitoring. Information is supposed to be collected on the physical condition of structures and canals, but there does not seem to be a framework for comparing this with previous years information to see whether the condition of the irrigation system is being maintained or improved over time.

M&E Systems

The problem is not a lack of systems for monitoring and evaluation. These have been developed in abundance. Elaborate systems have been created which may be feasible in a few favored sites, but are unworkable with the resources available in most areas. Elaborate information systems may simply not be implemented.

If implemented, information may be late, inaccurate and not linked to management decisions. Information on the more important objectives may be lost in a mass of data about less important items. There is a management saying that "what gets measured is what gets done." If monitoring systems have any use, it is as a means of directing the attention of irrigation managers, both farmers and agency staff, towards how to accomplish the goals of irrigation. If reforms are made so that participatory measurement of irrigation performance receives as much or more attention than measuring the physical and financial progress of construction, then much could be accomplished to improve performance.

Monitoring Key Tasks

One option to improve monitoring of irrigation O&M in participatory efforts would be to have a much clearer focus on indicators of performance in the key tasks of 1) equitably distributing water and 2) maintaining the condition of irrigation structures and canals. Both these can be measured fairly simply and objectively, using participatory methods. They are clearly linked to the impacts of irrigation in reducing risks and increasing yields, which help determine farmers' welfare and farmers' incentives to participate in irrigation management. While many factors affect yields, the key tasks under the control of irrigation staff and WUA, concern equity in water delivery and maintenance of the physical condition of structures and canals.

ON-FARM WATER MANAGEMENT DEVELOPMENT PROJECT OF IRRIGATION

The "Starter" project was intended to achieve a more efficient and effective use of water through increased cropping intensity and crop diversification while supporting self-reliance in O&M of farmer-managed irrigation systems. While implementation was mainly limited to a pilot project level, it is discussed in this chapter as an example of how agricultural development can be better integrated with irrigation development. This project was implemented from 1985 to 1992 by provincial agricultural services, with support from the Department of Agriculture and the Food and Agriculture Organization of the United Nations. Key elements of the project were field training, cost-sharing and training of agricultural officers.

FARMER FIELD TRAINING

Practical training was divided into a series of five sessions spread over two years. The first training developed a water management plan, preparing for system improvement and subsequent O&M. The main topics were mapping, problem identification, solution finding, water management action plan and water user organization plan. The farmers set priorities and provided ideas for the design of system improvements. This training was followed by improvements to irrigation structures, carried out by farmers. Later training sessions covered evaluation of the results of the water management plan (including system improvement), strengthening WUA organization, cropping patterns and agricultural development plans. Evaluation and adjustment of the plans followed at the end of each growing season.

Improvement of Irrigation Facilities

A subsidy worth about U.S. $1,000 per scheme was provided for construction materials. Farmers were expected to contribute labour, cash and materials of at least equal value. The project also provided limited funding for technical design. In many cases the subsidy for materials stimulated farmers to mobilize resources worth far more than the value of the subsidy. The fixed amount of subsidy per scheme greatly simplified implementation, created transparency and stimulated local contributions, since the amount of outside aid had been clearly specified from the start.

Staff Training

Agricultural officers received training to prepare them to support and guide water user associations to implement the on-farm water management programme. Skills were strengthened for data collection and analysis, problem solving, technical knowledge of on-farm water management, managing farmer training and organization and acting as mediators in FMIS development.

Impacts of Weirs Construction

Construction of weirs, canal lining and retaining walls benefited farmers through increased water availability. Construction of division boxes helped improve water distribution. Improvement in water availability and distribution increased cropping intensity and yields per hectare. These were combined with other agricultural improvements. According to project reports, production per hectare increased an estimated 20-30 per cent, raising farmer incomes about *Rp*. 400,000 per hectare per year. A comparative study of fifty turnover and Starter sites in West Java and West Nusa Tenggara found yield increases averaging about.25 tons per hectare, a lower but still substantial amount. Benefits from Starter project sites were as large as those at turnover programme sites, even though the Starter project improvements cost much less per hectare. However, Starter sites had more technical problems and seemed unlikely to last as long on average.

Incentives for Government

Government staff gained training in a variety of aspects of irrigation and agricultural development. There were only very limited funds available for the many visits to the field required by the participatory approach used in the project. Often staff had to obtain additional travel budget from other sources, or dig into their own pockets.

Several Issues

An evaluation by Rachbini raised several issues. WUA development focused more on the formality of organization, while traditional institutions continued to play the major role in irrigation management. There was no structure or unit to continue activities after the project. Trained officers usually went back to their routine assignments, which dealt with topics other than on-farm water management. The Starter project training for farmers included some diagnostic activities, to clarify local problems. More could have been done to increase the role of farmers in assessing what kind of training they need. Information and support services were fully subsidized, with no mechanism to enable farmers to share in the costs of extension services.

Influence on Other Projects

The Starter project demonstrated the potential for stimulating and mobilizing farmer participation. Some provinces, such as West Nusa Tenggara and East Java, used funds from other projects to apply the same approach on a wider scale. A technical assistance project funded by the Asian Development Bank (ADB), on Improvement of Farmer Managed Irrigation Systems (IFMIS), drew on methods from the Starter project, Turnover Programme and other sources to further synthesize a methodology for assisting FMIS. A project based on this methodology is scheduled to begin soon.

Village Irrigation Project

The Starter project and other efforts to assist farmer-managed irrigation systems have largely been overshadowed by the Village Irrigation Development Project. Launched in response to drought and rice production shortfalls in the 1990s, this crash programme emphasizes rapid construction of physical improvements, with construction organized by state-owned firms, using paid local labour. Provincial irrigation (functional) project units have managed the project, under the guidance of DGWRD. This project has funded works on irrigation systems covering many hundreds of thousands of hectares and is intended to cover a total of 1.6 million hectares identified as served by village irrigation systems. In some cases implementation has drawn on experience of participatory projects, to incorporate more consultation in design and more use of local labour and materials during construction than might have occurred otherwise. In general the emphasis has been on a crash programme to build physical infrastructure, without much attention to institutional or agricultural development.

CHARACTERISTICS OF IRRIGATION ASSOCIATIONS

The most distinctive characteristic of Irrigation Associations (*IA*) is that they are organizations of the people, for the benefit of the people. However, even though government participation is excluded by their very nature, most of these associations would never have existed without the promotion and encouragement of governments.

The Structure

From the organizational point of view an IA is made up of the following executive bodies.

- The General Assembly composed of all the farmers of the association. It is the highest authority. Its main function is to select their representatives (Board of Directors) and to approve or disapprove the management plans.
- *The Board of Directors:* This is the highest executive body. Its main function is to supervise and direct the execution of the work approved by the assembly and to prepare management plans annually.
- *The Manager's Office:* Is directly responsible for the execution of the mandate giver by the Board of Directors and for day-to-day work.
- *The Executing Units:* Are responsible for specific functions, such as operation, maintenance and administration.
- *The Irrigation Juries:* Some of these associations - especially the traditional ones - have their own juries to punish faults against the set rules and regulations. The jury is generally selected from distinguished members of the Board.

Supporting Agricultural Services

The very existence of a water management organization implies that a segregated organizational structure has been selected and that therefore the other necessary services are provided by other institutions. All too frequently this assumption is arrived at too quickly since the mere presence of extension, credit and other services in the project area does not mean that they are available to farmers in sufficient quantity to satisfy their specific needs. This is particularly important in the case of irrigation extension services which are a prerequisite to proper use of water and other agricultural inputs.

To ensure coordination with the agricultural services and to engage them actively in the affairs of the irrigation scheme, it is good practice to include representatives of such institutions on the Board of Directors of the association as special members. Alternatively, representatives of the IAs should be invited to the decision-making meetings of the concerned institutions at regional level or, where justified, at national level. In any case, it is important to assure the possibility of a dialogue between the institutions providing the services and the representatives of the farmers receiving them.

Main Characteristics of Irrigation

This includes their origins, main types (modern and traditional), legal character, water use rights of the association and its members, operation of their main functional bodies and financial responsibilities.

Reference is made here only to their most outstanding characteristics:

- Irrigation Associations (IAs) secure farmers' participation in decision-making for the irrigation schemes. The democratic process of selecting farmers' representatives guarantees such participation.

In a world where the people's desire for participation in public affairs is expanding quickly, securing such participation is most relevant.

- In this connection it may be opportune to raise the issue of the non-political nature of these associations. Their development has sometimes been handicapped because of the belief that these associations may become political and so be a powerful instrument of political leaders. Although in theory such a risk exists, practice has demonstrated that it rarely happens.
- IAs appear to be more effective administratively than Public Irrigation Schemes. Farmers are personally interested in their own organization and this sometimes permits them to do the same job cheaper and faster. Thus IAs reduce the need for a heavy public bureaucracy to run the irrigation schemes. The civil servants utilized in the development of an irrigation scheme can be moved to a new scheme when the first is completed, instead of remaining as permanent employees of the developed scheme.
- The recovery of water fees is more effectively carried out through IAs than through Public Irrigation Schemes. Experience shows that payment of water fees is higher among farmers of IAs than in those run by government officials.
- Rules and regulations are respected better since the punishment of faults in IAs is effective and quick. Most of the IAs have their own irrigation juries or regulations that are highly respected by members, so making any punishment effective.
- The relations between those distributing the water (water masters and water guards) and the farmers receiving it are much more friendly and cooperative than when the same job is undertaken by personnel of the administration. In the latter case, the water masters and ditchriders are seen as representatives of the government and not as part of the farmers' group.
- The IAs provide an excellent communication channel between the administration and the farmers. When they do not exist, communication tends to be one-sided from the top to the bottom.
- The Board of Directors and often the appointed Manager himself are farmers of the irrigation scheme and continuation in their jobs will depend largely on their performance. This is indeed an important incentive to carry out their jobs in the best possible way.

In spite of the fact that IAs are a highly desirable way of managing an irrigation scheme, they also have some limitations:

- The water distribution system utilized frequently leads to considerable operational water losses. Most of the IAs operate on the basis of a 'semi-demand' system, *i.e.* the farmer requests water and within a period of 3 to 7 days his request is granted. Although this system is perhaps the most desirable one from a social point of view, it is less efficient than the rotational system. This can be easily understood when one considers that to grant the request of a few farmers sometimes the whole canal irrigation network has to be kept in operation, with the corresponding water losses.

- Sometimes old water rights are an obstacle to efficient water distribution. In the course of their history many old irrigation associations have developed particular water rights which are incompatible with an efficient operation of the network. Such systems usually grant a fixed amount of water to the landowner and such a right can be sold or bought from others. This leads to all kinds of transactions - many of which bring social unrest - and subsequently to the need for constructing small but long ditches to bring the water to the buyer's land. Operational losses are bound to be high in this kind of water distribution system and consequently infiltration losses will be considerable owing to the complexity of the distribution.
- There is little capability for undertaking responsibilities outside the operation and maintenance field. Most of the people running an IA are from the farmers' community and are unlikely to have technical and managerial skills, thus tending to limit their sphere of competence to problems that they themselves can easily handle. So the involvement of IAs in farmers' training activities or water research investigations, monitoring of water quality problems, marketing and agricultural processing activities, etc. becomes problematic. However, it is possible for the IA to become more involved in activities other than those purely related to operation and maintenance problems by having qualified personnel. The Irrigation and Water Districts of the USA prove that under proper technical guidance and with sufficient financial resources, IAs can be very dynamic organizations with a wide sphere of competence.
- Long periods of considerable effort are required to get an IA established and working properly, even in countries like Spain where such institutions have centuries of tradition. The farmers' irrigation experience, educational level, the tradition of such associations in the region, the absence or presence of farmers' leaders are all factors that may influence the time required for the proper functioning of such organizations. In favourable conditions the IA can be operational a few years (5 or less) after completion of the irrigation works, while in less favourable conditions long periods of 15-20 years may be required.

Size of the Association

The question of the optimum size of the Association is important in determining its potential capability to undertake the related functions efficiently. Indeed, up to a certain size of irrigation scheme (about 5000 ha), the size of the Association is predetermined by the physical size of the scheme. In larger schemes, the possibility of subdividing them into irrigation sections, each having its own IA, or having a large association for the whole scheme, is frequently debated.

Large IAs have the advantages of greater economic potential and negotiating power. However, the larger the IA the more difficult becomes communication between the individual farmers and the executive body. In countries where the means of communication (mainly telephone) is effective and widespread in the rural areas, the formation of large IAs may be desirable (this is the case in the Irrigation and Water Districts of the USA), otherwise a federation of smaller associations may be preferable.

On the other hand, small IAs facilitate communications; but the administrative costs are greater and therefore place more of a burden on the farmers. From these discussions it appears that the optimum size for an irrigation association is neither too large nor too small: a size corresponding to 10000 ha (2000 - 4000 farmers) is perhaps a desirable target. This should be matched with the criterion that each IA should operate within the command area of hydraulically independent canals, as far as possible.

REFORMING WATER MANAGEMENT THROUGH ENVIRONMENTAL ETHICS

The reform of water policies is normally claimed to be the domain of lawyers, who write the laws, and economists, who apply economic logic to demonstrate the need for new laws and policies. Recent efforts to promote environmental flow as a core feature of sustainable water management utilize economic principles to provide the justification for environmental action, and bring in legal reforms as part of the solution. The case of the Santa Fe River, however, suggests that environmental economics (based on hydrology and ecological science) is limited in its influence.

The municipal water managers who control reservoir releases into the Santa Fe River are not interested in conducting a research study into the environmental economics of their management strategy. Local environmental NGOs such as the Santa Fe Watershed Association have a strong interest in environmental economics but lack the means to undertake the research. Water management institutes have shown little interest in becoming involved in the local concerns of the relatively small Santa Fe River, and would likely face criticisms from local and state authorities for interfering with local water decisions. In the absence of environmental economic analysis, the status quo management of the Santa Fe River can only be challenged conceptually, where, by definition, the status quo dominates. With no compelling case for applying environmental flow to the Santa Fe River, there is also no interest in developing a supportive legal framework. The existing laws already serve the dominant economic interests. In the eyes of Santa Fe's political establishment, there is no environmental problem; the only water problem is one of supply, and the concept of environmental flow would only exasperate the supply problem by catering to yet another consumer, the river.

Climate and Driver of Policy Change

Concerns about climate change and the security of Santa Fe's water supply offer a possible stimulus for new thinking about the management of river flows. The city's water supply derives from two sources;

1. The Santa Fe River (by impounding the flow) and its associated aquifer and
2. The Rio Grande River (by river diversion and pipe) and its associated aquifer. The anticipated effects of climate change include
   - Relatively less snow and more rain, resulting in less effective water storage in the form of snowpack in the upper watershed, and
   - More severe and more frequent droughts and floods.

The availability of surface water will almost certainly be more variable and risky, even as there will be a need for greater storage capacity to withstand the anticipated multi-year droughts. Aquifer storage will become a more valuable part of the water supply strategy and there will be a simultaneously greater need for empty reservoir capacity as flood buffer. For the Santa Fe River, the importance of river flows for recharge might prompt some rethinking of reservoir releases.

A scenario of environmental flows that recharge the aquifer, while helping fortify the riparian ecology (vegetation and river channel morphology) could result in a healthier, more resilient, "living river". Is such an outcome possible within the "water-as-resource" utilitarian ethos of Santa Fe's water managers? There will certainly be pressure to respond to climate change by increasing surface storage capacity rather than allowing any river water to infiltrate into the aquifer.

The priority of surface storage over aquifer storage has been the convention for the past half-century. With no value accorded to riparian health, the additional environmental benefits from flows that infiltrate water into the aquifer would not be counted, and the comparison would be made on storage and recovery criteria alone: How much water can be recovered and at what cost, if the water is allowed to infiltrate into the aquifer rather than be stored in the reservoirs? With the higher temperatures and greater evaporation under climate change scenarios, coupled with the increased probabilities for extended droughts that would in any case render surface reservoirs useless, maintaining instream "environmental" flows might be seen as economically preferable even without any environmental considerations. Perhaps Santa Fe's water managers will be induced to take the environmentally enlightened option in spite of themselves. Climate change will put new stress on water management policies everywhere, and the same types of choices confronting Santa Fe's water managers will become the norm. Faced with greater uncertainty and more extreme floods and droughts, will we try to exert even greater control over unruly rivers, or will we be induced to look for more ecological approaches? Continuing with the conventional command-and-control paradigm would inflict even more damage on already degraded water ecosystems. The urgency of responding to climate change offers an opportunity to take corrective action, to re-orient water policies to operate with the natural ecology and to take advantage of "nature's infrastructure".

Crisis of Water Management

Can water management embrace ecology without a crisis? Can we make preemptive reforms to water policies that embrace Nature as having a right to exist and a value above and beyond the ecosystem services that benefit humans? Based on my experience in trying to effect policy reforms in Santa Fe, my sense is that environmental ethics is not going to become a compelling rallying cry for reforming water management. Cultural values and ethics are inherently resistant to change; that is how cultures are maintained over time. Rather than attempting low probability heroics with a frontal attack on well entrenched values, a more effective strategy might be to look for ways that the existing value system might be used to justify environmentally important measures that could lead to larger reforms. An approach of seeking to change behaviours within the umbrella of existing values can start a multiplier effect that can influence values gradually and indirectly. The role of the crisis (*e.g.*, climate change) can be to stimulate the initial behavioural change.

In the Santa Fe case, the need for multi-year carryover storage that is resistant to evaporation could serve as a stimulus for experimenting with "aquifer storage and recovery" (ASR), using instream flow in the Santa Fe River. By appealing to the intuitive logic of storing water underground for long periods of time, a living Santa Fe River becomes not an extremist environmental idea, but a delivery mechanism for storing large amounts of water as drought insurance. As the flowing river and growing aquifer become accepted as a normal aspect of safeguarding long-term (sustainable) water supply, the value placed on a flowing river could become an internalized aspect of the local culture.

Healthy Rivers as a Management Goal

When a healthy river is seen as a management objective for its own sake, utilitarian objectives are forced into an accommodation with sustainability.

Under such an approach, the City of Santa Fe would no longer be able to impound the entire flow of the Santa Fe River merely because this is an easy way to obtain water for the city supply. With the health of the Santa Fe River as a central management objective, the needs of urban water supply would have to be met in ways that did not endanger basic ecological functions. The dams would be operated as part of a conjunctive strategy with groundwater management, demand management (conservation), rainwater harvesting, and water reuse.

Setting the health of the river as a priority is a precondition for garnering the political support for reforming the legal framework that would support the implementation of ecological water management. But who can set new priorities without having that political support to begin with? This is the role of the "invisible hand" of cultural values. Like Adam Smith's unseen hand that allows economic markets to function efficiently, the unseen hand of culture can also appear sui generis under the right conditions, to protect the interests of Nature through the interplay of individual interests. The evidence that environmental protections are natural outcomes of culture is found in the sacred respect by which rivers, lakes, streams, and springs are held in every indigenous culture, present and past.

Of course, we don't have time to deprogram our current culture and to uncover the forgotten indigenous heritage of spiritual water values. But perhaps science can once again serve as a substitute for religion and help extricate our society from the unsustainable mess we've created. Applying principles from ecology (including hydrology) and social science (including economics and anthropology), we have the tools to recover our lost innocence and to learn to live in compliance with the natural laws governing water ecosystems. Climate change is prompting a search for new approaches; perhaps we should try an approach endorsed by Nature herself.

THE CONVENTIONAL WATER MANAGEMENT PARADIGM

The conventional paradigm for managing water during the 19th and most of the 20th centuries was based on the principle of "command-and-control." Dams were built to impound rivers, and release the water on command, for power generation, or into canals for irrigation, industry, or urban water supply. River channels were straightened and deepened; levees kept rivers contained and away from their natural floodplains. Urban runoff from rain and snow which fell on roofs, streets, and parking lots, was channelled and funnelled through drains and sewers to get back to the nearest river as quickly as possible.

The principle of "command and control" was, and still is, intricately intertwined with the principle of "beneficial use" and most particularly, economically beneficial use. Under the conventional paradigm, water needs to be tightly controlled in order to direct its service to a "stream" of economic benefits. If water is not providing economic benefits for somebody, it is not being put to beneficial use. In the Western part of the United States, a third principle also comes into play, that of "prior appropriation." Based on the idea of mineral prospecting, this principle grants permanent water rights to the first person to extract water for an economic use. The practical implication of this principle is that any water flowing in a stream that is not already claimed by someone else (for a particular economic use) can be claimed by a newcomer, so long as that newcomer can demonstrate that the water will be put to an economic application (*e.g.*, used in a factory, or for irrigation, or for municipal drinking water, etc.).

The idea of putting water to beneficial use is a very old concept. The Sri Lankan King Parakramabahu the Great (1153-1186 CE) famously said "No drop of water that falls from the sky should be allowed to flow to the sea unused." He implemented his own advice by building reservoirs and canals to command and control the water. From a physical, engineering perspective, there are remarkable similarities between ancient Sri Lankan water management practices and the practices of today.

During the construction of a sluice gate for a new reservoir in the Mahaweli development project, an ancient sluice gate was discovered on the same location, from an earlier reservoir. What was different, aside from the contrast of stone vs. concrete and steel, was that the ancient stone sluice gate was constructed in the form of a seven headed cobra deity. By honouring the water god, the ancient engineers were acknowledging the "prior appropriation" right of the river spirit, and seeking both permission and protection. Conventional water management shares the "command-and-control" approach of medieval Sri Lanka, but without the spiritual underpinnings. This view of water and rivers as inert material resources possessing neither consciousness, nor spirit, nor any inherent rights that need to be respected comprises a key principle of the conventional paradigm: Water management is a secular undertaking; rivers and lakes are regarded simply as accumulations of inert matter.

Historians have traced the development of the conventional water management paradigm to European and American political and cultural dynamics. Donald Worster (1985) provides an analysis of water development as an expression of frontier expansion into the American West during the 19th and early 20th centuries. David Blackbourn (2006) takes a similar approach in showing surprising parallels in the evolution of water engineering and the development of modern Germany over the past two centuries.

Environmentalists have waged a long, and generally losing war against the conventional water management paradigm and the destruction of unique natural landscapes. One of the first major battles was John Muir's effort to protect his beloved Hetch Hetchy Valley in the early 1900s. Originally included as part of Yosemite National Park in California, the US government withdrew its protected status to allow construction of a reservoir for the growing city of San Francisco. Muir, founder of the Sierra Club, argued that there were many other potential sources of water for San Francisco, but only one Hetch Hetchy Valley. His impassioned pleas fell on deaf ears: "Dam Hetch Hetchy! As well dam for water-tanks the people's cathedrals and churches, for no holier temple has ever been consecrated by the heart of man".

The Greening of Conventional Water Management

The sad history of environmental failures to preserve river ecosystems in the United States is documented in Marc Reisner's Cadillac Desert. Invariably, pleas to leave rivers intact for posterity were no match for money and politics. While countless individual battles were lost, however, the overall campaign to protect rivers from unchecked dam development did have some success.

Overtapped Oasis, Reisner (1990) reviews the history of Western water development up to that time, and notes that the bleak picture he had painted only a few years ago in Cadillac Desert was already looking better. The era of new dam construction in the US had come to a close, partly because the best locations for dams already had them, but also because the environmental costs were being more carefully assessed through economic cost-benefit analyses. Today, it is clear that Reisner's hope for more environmental awareness about water management is being at least partly realized.

Restoration projects aimed at mitigating the ecological damage from unfortunate construction projects of the past have become routine. The conventions of water management are shifting towards more emphasis on sustainability and ecosystem services, but there is far more rhetoric than action in adopting ecological principles. Water management continues to be defined as a series of physical challenges to be solved through engineering rather than adaptive management. A fundamentally new approach is needed, and fortunately, has already been developed by ecologists, biologists, and resource economists.

Ecological Water Management

The ecological water management model is predicated on sustainable ecosystems. A fundamental feature of ecological water management, in contrast to "command-and-control" approaches, is a commitment to environmental flow. An environmental flow is the natural water regime of a river, wetland or coastal zone which maintains the ecosystem. A minimum environmental flow is the smallest amount of water required at any given time to allow the ecosystem to function. Environmental flows provide critical contributions to both river health and ultimately to economic development, ensuring the continued availability of the many benefits that healthy river and groundwater systems ring to society.

Since the 1990s, the concept of environmental flows has been gradually incorporated into water laws from Europe to South Africa to Australia. The South African National Water Act adopted in 1998 granted water resources the status of public goods, under state control; the national government is the custodian of the water resources and its powers are exercised as a public trust. It has the responsibility for the equitable allocation and usage of water and the transfer of water between catchments. The Act establishes a 'reserve 'consisting of an unallocated portion of water that is not subject to competition with other water uses. It refers to both quality and quantity of water and has two segments: the basic human need reserve and the ecological reserve. The first one refers to the amount of water for drinking, food and personal hygiene and the second one to the amount of water required to protect the aquatic ecosystems.

In Europe, the Water Framework Directive, enacted in 2001, requires sustainable water levels and flows to maintain or restore riparian habitats. Three groups of quality standards (biological, hydromorphological and physico-chemical) are identified as necessary to a healthy ecological status. Member States are required to achieve good surface water and groundwater status by 2015, which refers to ecological and chemical status of surface waters, and chemical quality and quantitative status for groundwater, with rates of abstraction sustainable over the long term.

Other dimensions of ecological water management include the morphology of river channels (which should be meandering with functioning floodplains and natural vegetation) storm water management (capturing, slowing, and spreading water so it can infiltrates into the soil, and water use whether for domestic, industrial, or agricultural purposes. In each domain of water management there are opportunities to utilize ecological principles to capture, store, purify, and reuse water. These complex spheres provide literally endless ways of impacting water ecosystems, which can be helpful, benign, or harmful.

The Water Response to Climate Change

As people accept that climate change is real and here to stay, they are likely to realize that while reducing greenhouse gas emissions is all about energy, adapting to climate change will be all about water.

- Frank Rijsberman, former Director General, International Water Management Institute.

Climate change is adding a new dimension to water management. Within the conventional water management paradigm, climate change adds urgency to more and bigger technical fixes. The inclination is to go "back to the future" of command-and-control approaches and re-engineer a solution to climate change: Build more dams to store more water, more pipelines to cross basin boundaries and more pumping to tap ever deeper sources of groundwater.

Environmental voices within the conventional paradigm call for greater water conservation efforts and smarter "conjunctive use" strategies to rely on surface water during wet years and let the groundwater build up for the dry years. Along with the calls for new and greener technologies, there is also an emerging consensus among environmental groups that restoring ecological health to rivers, lakes, and groundwater is essential to provide the resilience that will be needed to weather the anticipated greater swings of longer droughts and bigger floods. Our best hope as humans, according to this approach, is to rely on "Nature's own infrastructure". If we can keep Nature functioning, we can survive the tribulations of climate change and prevent any further damage to the rivers, lakes and aquifers which our still growing population will depend on.

The dilemma of finding consensus around the best response to climate change stems from the fundamental incompatibility between the conventional and ecological water management paradigms. Building more dams on already dewatered rivers (as often proposed in the conventional paradigm) will further damage the very ecosystems that the ecological paradigm is trying to restore. Proponents of an ecological approach have science on their side. There is a clear trend within the scientific community in favour of ecological solutions which lend themselves to unanticipated synergies, rather than conventional responses which often result in unexpected collateral damage. But proponents of the conventional paradigm still have politics on their side, as we will see in the cast of the Santa Fe River in New Mexico.

COMPETING PARADIGMS OF WATER

The Santa Fe River emerges from high (3,500m) mountains to the East of Santa Fe City and flows 75kms in a Westerly direction to join the Rio Grande. Two reservoirs in the mountains impound the entire flow of the river for City reservoirs, providing about half the water the city uses each year. The other half of the water comes from deep wells in and near the city, including from the Santa Fe River aquifer. Santa Fe's water policies are based on 19th Century Western US water law and can be summarized in the phrase, "First in time, first in right." Earlier claims to water trump later claims, other things being equal. The most important of these other things is putting the water to "beneficial use," which means an economically productive purpose. Omitted from the law is any consideration of the water resource itself. Neither the rights of nature in general or the rights of a river in particular, are accorded a seat at the legal table.

History of Water Use Along the Santa Fe River

Indigenous Pueblo Indian tribes were already using the Santa Fe River for irrigation when the Spanish arrived in the late 1500s. With the establishment of Santa Fe as a provincial capitol in 1610, agricultural use of water intensified. The Santa Fe River provided water to a growing network of Spanish canals (acequias) which provided the food for the growing settlement. More that 30 acequias were established, irrigating roughly 800 hectares of farmland, and diverting so much water that stretches of the river were dry during the summer months.

Based on the cultural values underlying contemporary acequia agriculture, it seems safe to suggest that the colonial Spanish viewed the river primarily as a means of cultivating a secure food supply in this semi-arid environment. The rights of nature were implicitly assumed: "The tacit, underlying premise [of acequia irrigators] is that all living creatures have a right to water". Another core belief related to water was that it should be shared: "The principle of water sharing belongs to a larger moral economy that promotes cooperative economic behavior through inculcating the core value of respecto and gendered norms of personal comportment".

The 19th Century saw major changes in the cultural orientation towards the river. The United States annexed the territory of New Mexico in 1848. Thirty years later, the first dam was built on the river, to provide municipal water for the newly established Santa Fe Water Company. Initially, the Water Company stored less than 10 per cent of the river's flow for its customers, but by the mid 1900s, that proportion increased to nearly 50 per cent. From now on, the river's water would be the basis for continued urban expansion, rather than agricultural production.

Today the river is operated with the objective of storing as much water as possible in the reservoirs. The rights to the river's water were extracted from the acequia farmers through legal maneuvers by the private water company, which later sold the rights to the City of Santa Fe. The city's rights comprise about 85 per cent of the river's average flow (which is highly variable). An additional 5 per cent is owned by the heirs of the early farmers (now used for urban gardens) and the remaining 10 per cent is unallocated, usually spilling from the reservoirs when the mountain snowpack melts in late Spring.

The reservoir dams are operated by City engineers. Water is normally released from the dams in anticipation of Spring floods, and (in a recent policy evolution) for aesthetic purposes during the summer months. During the rest of the year, the river is a dry, heavily eroded ditch. The policy of keeping the river intentionally dry earned it the designation as "America's Most Endangered River" in 2007.

Partly in response to this negative publicity, the City government, which controls the reservoirs, shifted its policy to allow for summer water releases into the river provided the reservoirs are ¾ full rather than completely full. A year-round minimum environmental flow, however, as required in other countries, is not under serious consideration. The City's policy remains one of prioritizing storage at the expense of flow. In the prevailing view of the

Santa Fe's municipal government (whose view matters most, since the municipality owns the river's water rights), the primary and almost exclusive purpose of the river is to provide water for the city's municipal water supply, thereby supporting existing homes and businesses plus future expansion. The idea that the river itself has a rightful claim to some of its water lies outside the prevailing paradigm, which is based on a legal perspective. If you have a right to the water, and if you are complying with the legal requirement that the water be put to "beneficial use", there is no legal restriction on water withdrawals.

Pealing back the legal and economic veneers we can expose a layer of ethics that is otherwise obscured. The willingness to accept the City's legal entitlement as an adequate justification for dewatering the river implies consent with the principle of water as a commodity that can be freely owned and traded. The environmental ethic being expressed is that water is a natural resource that can (and should) be utilized for economically productive purposes. A corollary to the "water-as-resource" principle is that the ecological context of the water (river, lake, aquifer) lies outside the concern of resource management. In terms of environmental ethics, the municipal government of Santa Fe is not being un-ethical (violating the ethical code) but merely a-ethical (lacking an ethical code).

These two principles together explain how the city water managers, who are well educated professionals and generally aware of ecological science, can take the position that the health of the dewatered Santa Fe River is not a problem. By not subscribing to the principle that the river's health has an ethical importance, the test for determining whether river health deserves to be considered as a management will be based solely on economic considerations of managing the water for greatest economic benefit. This management objective is implied in the legal terms defining the conditions for meeting the legal requirements of water-as-property, namely that the water be applied to some "beneficial" activity.

This is commonly interpreted to mean domestic water, landscape or crop irrigation, or any industrial purpose. Local water attorneys are divided in their opinion as to whether leaving water in the river for environmental benefit can qualify as a "beneficial" use under the law. Water managers in New Mexico routinely express their professional mandate as making water available to the water rights holders, and specifically to "senior" rights holders, since under the law, junior water rights can be claimed only after the senior rights holders have received their water allocation.

During a water shortage, for example, the owners of the oldest recorded water rights (based on the date that a claim was filed in the state courts) have a legal right to their full amount of water, as if it were a normal or wet year, before the more junior right holder's rights get activated. In practice, there is usually a sharing of the shortages but there is no legal requirement to share.

If a state water manager were to express the objective of making water available for the health of the riparian ecosystems, he would almost certainly be removed from office for advocating a violation of state water law.

Ecological Water Management

The blatant disregard for the ecological health of the Santa Fe River, which flows through the middle of a state capitol which is also one of the most popular tourist destinations in the country, has stimulated local efforts to restore a "living river". An initiative from within the city's governing council led to a comprehensive Santa Fe River Corridor Master Plan in 1995 which endorsed the objective of a permanently flowing Santa Fe River, but without any stipulations for how this might be accomplished. The reasons given for why this objective was important were couched in broadly utilitarian values:

- We advocate surface and groundwater management that balances human use with natural resource protection. We encourage government and civic leaders to place high priority on sustaining seasonal stream flow in the river, yielding hydrologic, recreational, aesthetic, and environmental value to the community. We are committed to safeguarding the long-term integrity of the river and the entire watershed.

Pieces of this agenda have been incorporated into City policies. The municipal government has officially endorsed the goal of a living river, including a 20 per cent allocation for river flow, but subject to completion of a new pipeline to secure an additional source of water, and even then only in normal or wet years.

Under dry conditions (less than 80 per cent of normal precipitation) the river flow would be shut off in favour of reservoir storage. The precise amount of water necessary for an effective environmental flow is a matter of debate even among environmentalists, but the City's planning provision for shutting off the river completely during moderately dry years clearly lies outside the concept of "environmental flow". Without the support of legal requirements establishing minimum flows or other ecological standards, river advocates such as the Santa Fe Watershed Association rely on persuasion and community interest.

The water management approach advocated by the Santa Fe Watershed Association include the following features:

- Commit a permanent minimum environmental flow to the river, which would vary according to reservoir levels but always be greater than zero (even during droughts when the reservoirs may be empty).
- Manage the reservoirs to release the Spring runoff over a period of months rather than uncontrolled spills;
- Provide financial incentives for rooftop rain catchment systems;

- Initiate water conservation campaign and incentives, and allocate a portion of conserved water to be returned to the river;
- Enlist the help of private well owners (who account for ca. 10 per cent of total urban water use) through buy-back arrangements to free up water for the river
- Capture storm flows from streets and parking lots and infiltrate into the shallow groundwater and the river.
- Increase groundwater pumping capacity for contingency during droughts

City officials accept the general concept that a flowing water is a good thing, but they stop short of endorsing the principle of environmental flow. They argue that there is not currently "enough" water available for the river. Only after the total supply is increased can the "new" demand of water for the river be accommodated, and then only under normal or above-normal precipitation years. As a City official explained to me, "The river has to live within its means." In the eyes of the City water managers, the river is a water "consumer" which competes with other consumptive demands: drinking water, irrigated gardens and lawns, golf courses, and new housing and office buildings. Each of these existing demands needs to be "balanced" and this balancing is the art of water management. The environmentalist demand for enough water for the river to support its ecological function is viewed as an unfair concession to the environmentalists. Instead, a balanced and "prudent" approach is required.

The Problem: Cultural Values. Why is the Santa Fe River considered a competing "consumer" with no claim to its own water, when in South Africa a minimum water flow is legally protected as an "ecological reserve"? The problem is not lack of scientific knowledge or expertise. The problem lies in the realm of cultural values, and specifically values related to the role of nature *i.e.*, environmental ethics. In Santa Fe, the river is considered an inanimate thing, a physical channel through which water sometimes flows. It is not the subject of ethical concern, much less religious veneration. The river is not even seen as a provider of water, which now comes from dams, wells, and pipes. Nature has ceased to have direct relevance to the issues of water management. The river has become a "community amenity" like a park or a garden. Nature has been tamed and can now be managed through technologies, and allocated according to a prudent and balanced assessment of society's competing water demands.

BACKGROUND OF CONVENTIONAL SYSTEMS OF WATER MANAGEMENT

Conventional systems of water management have been considered one of the 20th Century's greatest public health accomplishments. Large networks of water and sewer lines and treatment plants brought clean water into the cities and transported away disease-carrying sewage. With passage of the 1972 Clean Water Act, the nation also established a goal of fishable, swimmable surface waters and funded and enforced increasingly stringent regulatory standards for the partial removal of pollutants by treatment plants. The 1974 Safe Drinking Water Act focused on setting up engineering barriers to protect drinking water from pathogens, but in doing so ramped up the cost of providing water treated to a potable standard that is also used for many non-potable uses.

Municipal utility management was the norm. Rural towns with private wells and septic systems were expected over time to build similar public water and sewer systems as well. In recent years, however, a concern has been growing that this "paradigm" of big-pipe water management is not sustainable, both from a natural resource and a financial perspective. The appropriation of huge volumes of water from ecosystems and the release of polluted effluent into rivers and the oceans have been increasingly disruptive to ecosystems. Regulations have required removal of some pollutants, but relied on "dilution" in receiving water as well.

Signs of stress are seen in falling groundwater levels and decreasing dry-weather stream flows (and unnatural flow increases during wet weather), eutrophication of lakes and estuaries, disappearance of wetlands, dead zones in coastal areas, and other changes in hydrological functions. Climate change is expected to exacerbate patterns of droughts and heavy rainfalls, putting both water supplies and flood control measures at risk. Reductions in evapotranspiration are being studied as potentially significant contributors to global warming.

Drinking water systems lose huge amounts of water (a US average of 20 per cent) from their distribution pipes, treatment technologies were not designed to minimize emerging biological and chemical contaminants, and treating all water to new and more stringent standards is both increasingly difficult and expensive and, except for the small amount of water needed for potable uses, is extremely wasteful of energy, chemicals and money. Most cities and towns have been unwilling to charge ratepayers the full cost of repairing and replacing the existing inadequate infrastructure, and so collapsing pipes and breakdowns in treatment plants have become more frequent, while innovation is generally off the radar screen.

The 2007 Baltimore Charter for Sustainable Water Systems suggests an alternative approach to water management that "mimics and works with nature". Natural systems create an abundance of value and diversity, where species cooperate and one species' waste is another species' resource. Introduced into these naturally-balancing ecosystems has been the highly-disruptive human extraction and use of resources of the industrial era. The genius of science and design in the 21st Century is, in contrast, the discovery of "smart, clean, and green" ways to capture the value of resources. "Smart" because they unlock the complex designs of nature and use information and signalling to achieve efficiencies. "Clean" because they capture and use resources and methods that don't involve significant externalities in extraction or disposal. And, "green" because they rely on vegetation, and in the process begin to restore the natural ecosystem and its wide and deep benefits.

In practical engineering terms, this new design model includes: provision of potable-quality water for drinking water and direct human contact purposes only; prevention of pollution before it gets into the waste stream (including the wholesale re-engineering of some products through green chemistry to mitigate or eliminate ecological damage); reduction of energy needs by avoiding the pumping and long-distance transport of water and wastewater; wastewater recycling and non-potable, "fit for purpose" reuse instead of disposal; rainfall harvesting and reuse to supplement water supplies; energy and nutrient recovery from wastewater; habitat and natural system restoration; re-vegetation to restore evapotranspiration capacity; and restoration of green infrastructure in urban areas to help beautify cities and revitalize neighbourhoods.

A birds-eye view of the new infrastructure would reveal "networks" of decentralized and repurposed and at times hybridized, systems. Some of the innovative treatment and resource recovery technologies would be "embedded" in subdivisions, apartment complexes, or individual homes and offices. Other functions would be taken over by vegetative "green infrastructure", such as green roofs and walls, trees, and swales along roads, and restored streams, riparian areas, and wetlands. Water and sewer lines might be slip-lined and repurposed for potable or reclaimed water, water storage and distribution, and heat recovery. Monitoring and control technologies would be key elements in managing these systems and in protecting public health and the environment.

These engineered and green networks mimic the natural systems of nodes and links in nature, where water both recycles and supports life at a local scale, but also is a linkage and transport mechanism across a landscape and into the atmosphere. These systems in cities and towns can cost less to provide water and sanitation services than current approaches and can also add significant benefits in terms of air quality, energy savings and production, recreation, beauty and aesthetics, increased property values, and jobs. Innovative pricing, incentives, and new performance-based regulatory mechanisms will be required to ensure that these sustainable practices are adopted and that the remaining watershed and global "externalities" are also addressed by developers, homeowners, and municipalities.

The analogy for such "network" design that is increasingly-known to the public is in the energy sector, where a shift to a distributed and efficient network that relies in significant part on clean natural system services is underway. The existing power grids use large electrical networks and power plants, as well as oil and gas pipelines, that deliver energy to homes and businesses at subsidized rates and produce large externalities in air pollution and greenhouse gas emissions. In contrast, the new grid will incorporate clean natural sources of energy, including wind, hydroelectric, and solar, at distributed locations and will encourage more energy-efficient building designs. Metering and incentives for peak generation and off-peak use of energy are the equivalents of "fit for the purpose" water provision. "Smart grids" also rely on sophisticated information and control systems. Both energy and water networks require supportive financial incentives and regulatory mandates.

A few leading-edge infrastructure experts are now suggesting that these "networks" of engineered and green energy and water systems need to be integrated and also be co-engineered with transportation, solid waste, buildings, and other urban infrastructure management. The lessons of nature are that such integration will lead to significant synergies of design, cost-savings, and an abundance of positive benefits for society. For example, an "eco-block" incorporating architectural innovations, wind and solar power, green roof and wall cooling, rainwater harvesting, water reuse and energy recovery, and nutrient recycling into community gardens, can be nearly "off-the-grid" in both energy and water, and can be located at transportation "hubs". These new designs of infrastructure may cost less in dollars and will both improve the quality of life in urban communities and begin to protect and restore the ecological Commons.

Paralleling the shift in technologies will be a shift in the institutions and markets for resource management. Municipal utilities were appropriate for each single-service "monopoly" centralized system in water, stormwater, and wastewater. But embedded and green infrastructure "nodes" in homes, subdivisions, and commercial establishments engage a wide range of private firms, non-profit groups, and other city agencies (such as parks and recreation, housing, job training, etc.), and the developer and property-owner will have many more choices for technologies and design and ongoing maintenance services.

Municipalities and other local governments will have more complex and highly-productive new roles in coordinating municipal utilities and agencies internally and in overseeing the new private and non-profit sector externally through ordinances, incentives, education, and inspections. Finally, the solutions to water management in the 21st Century will require a high level of interdisciplinary collaboration and broad public engagement. Here also, nature serves as a model for the benefits of collaboration and cooperation, or social capital, in society, as opposed to the specialization and hyper-individualism of the 20th Century. Networks of conversations and pilot projects will serve as the foundation for creative invention and enhancement of the "Common Wealth".

Full-Cost Pricing and Planning

This project will focus on the key driving elements in the emergence of the 21st Century water paradigm, which began to surface in the 2005-2006 workshop series convened by the Coalition for Alternative Wastewater Treatment.

The description at that time of a problem statement for this Project included the following language:

- One of the most significant barriers to the full and appropriate use of decentralized water resource systems is the failure to consider externalities of central system approaches. Siloed utilities and individual communities are allowed, subsidized, or forced by court decrees to proceed with highly-disruptive projects that cumulatively cost more and create greater environmental damage than an approach which both decentralizes and integrates various water resource sectors across a region.
- Examples of externalities that are not normally internalized to projects include: construction of water lines, followed by increased use of water that is tied to uncontrolled growth and an increased number of malfunctioning existing septic systems, and then the pressure to build sewers and large treatment facilities to deal with growth and existing failures of septic systems, followed by even more increases in development to help pay for these systems, and substantially increased stormwater runoff that releases pathogens and nutrients into surface waters, and sanitary and storm sewers that drain groundwater aquifers and exacerbate water supply problems.
- During the fall, 2005 workshops, Garrett Hardin's "Tragedy of the Commons" was discussed as a framework for understanding this problem. Possible solutions could include:
  - Pricing – true costs to customers
  - Conditions for grants and loans – Integrated Water Resource Plans
  - Permit requirements – NPDES, UIC, and others

Two of the six high-priority, short-term projects recommended by participants pertained to:

1. Research on full monetary and non-monetary benefits and costs of soft and hard path approaches, and pricing or other mechanisms to better align local decisions with long-run environmental and economic sustainability
2. Exploration of how to tie federal subsidies and permits to an integrated water supply and water quality plan in a watershed.

Updated Problem Statement

In the intervening three years, a series of water crises, Congressional hearings, and other conversations have expanded and deepened this topi considerably. The core insight remains – costs can be incurred or opportunities foregone when decisions are made by individual utilities, homeowners, developers, companies, municipalities, and other parties to the sector that only reflect short-term and narrowly-defined self-interest in minimizing costs and/ or maximizing profits.

Hardin's model was based on the over-grazing of open-access, community land (the British village green or "Commons"). The individual herdsman had no incentive to cut back the grazing of his own cows, since others would just take his place. But, cumulatively and over the long-term, the land would degrade and collapse in its capacity to support grazing for local herdsmen. Solutions to this "Tragedy" were to develop government regulations that restrict use or privatized the land. The costs of over-use would be directly experienced and "internalized" by each herdsman, who could over time capture the benefits of practices that protect ecosystem function. Later critics of Hardin pointed out that community-based management systems could also guard against over-use.

In water and water-related sectors, a concern is that a global "Tragedy of the Commons" may be on the horizon as well. As Jared Diamond has pointed out, past civilizations have turned lush landscapes into deserts through profligate use and mismanagement of water. The new literature on "resiliency" suggests there are tipping points in nature, where ecosystem collapse is difficult if not impossible to reverse. For example, a lake can absorb and store some nutrient pollution, but at some point the lake "dies", and the eutrophied water body can't be cleaned and revitalized at a reasonable cost. There are similar concerns for tipping points in the capabilities of seas and oceans to absorb carbon and acids.

Water is connected to so many ecosystem and societal functions and services, from the elemental to the spiritual, that the phrase "water is at the heart of all life." is often used to describe those relationships. The first goal of this Project is to examine water's many connections to the "Commons" and to identify those links, in particular, that may be stressed in the future to the point of collapse.

These may include depleted water supplies for potable, commercial and industrial uses, desertification (drying and degradation of soils), redistribution of phosphate stores from where needed for growth of plants to water bodies causing choking algal growth, toxic overloads of the receiving environment, unnecessary energy use to move and treat waters, and disruption of evaporation cycles that result in climate change, and others. There are numerous other impacts on the Commons that may fall short of "collapse", but that nevertheless represent a degraded quality of life on Earth. "Commons" is a useful term to describe the space in which natural organisms, including humans, interact. There is an ecosystem web of life with complex interdependencies of species, natural resources, energy, and soils and water.

There are also complex interdependencies of people in human societies and economies, which in turn rely on nature's services for their survival. The Commons can be used to describe that space of interactions and interdependencies that affect the well-being of all the individual participants, both human and non-human.

This is a particularly interesting and critical moment in history, when both the ecological and societal Commons are threatened. Climate and other natural systems are under increasing stress globally, but so is the economy. New discussions are being held about the need to restructure relationships in the Commons so that ecological, social, and economic systems are restored to health.

Advocates of a "green economy" assert that governance reoriented to ecosystem protection and restoration, in particular greenhouse gas reductions, can also be good for society and the economy. Climate change can be slowed, while at the same time "green jobs" are created, air pollution is lessened, and costs of energy are reduced. Advocates for a new paradigm in water management make similar arguments.

President Barack Obama and others looking at a fundamental restructuring of the economy hasten to state that these conversations are not about "socialism" or "central planning", but rather about identifying the disincentives and "market failures" in the current economy that need to be addressed. Properly-functioning markets are an efficient means of resource allocation, but participants need the proper information, signals, and incentives for internalizing and incorporating actions and behaviours that benefit the Commons.

The "sweet spot" is when all sides benefit, as for example, when education enhances the earning power of the individual as well as the productivity of the overall economy. Another example promoted in the fiscal stimulus bill is weatherization of homes, where the homeowner sees lower energy costs, but jobs are also created, and ecosystem damage is minimized. The new Administration also recognizes that social and economic development will depend on a better integration and coordination of services, particularly at the local level. In urban policy, for example, President Obama has espoused an approach that would integrate education, health, job creation, infrastructure, green space, and housing renewal programmes in targeted neighbourhoods.

One of the effects of this new strategy may be to grant more authority to local governments and to loosen the rigidity of specialized mandates from the state and federal levels. Finally, there is an impulse in society for restoring public spaces and social interactions, whether it be in farmer's markets, revitalized Main Streets, information networks, democracy forums, and others. These three strategies, a restructuring of markets, an integration of services at the local level, and expanded public opportunities for creative invention, will support the emergence of a complex natural systems-based approach in water management, as well.

IRRIGATION WATER QUALITY CRITERIA

Salt-affected soils develop from a wide range of factors including: soil type, field slope and drainage, irrigation system type and management, fertilizer and manuring practices, and other soil and water management practices. In Colorado, perhaps the most critical factor in predicting, managing, and reducing salt-affected soils is the quality of irrigation water being used. Besides affecting crop yield and soil physical conditions, irrigation water quality can affect fertility needs, irrigation system performance and longevity, and how the water can be applied. Therefore, knowledge of irrigation water quality is critical to understanding what management changes are necessary for long-term productivity.

IRRIGATION WATER QUALITY CRITERIA

Soil scientists use the following categories to describe irrigation water effects on crop production and soil quality:

- Salinity hazard - total soluble salt content
- Sodium hazard - relative proportion of sodium to calcium and magnesium ions
- pH - acid or basic
- Alkalinity - carbonate and bicarbonate
- *Specific ions*: chloride, sulfate, boron, and nitrate.

Another potential irrigation water quality impairment that may affect suitability for cropping systems is microbial pathogens.

Table. General Guidelines for Salinity Hazard of Irrigation Water based Upon Conductivity.

Limitations for use	Electrical Conductivity (dS/m)*
None	≤0.75
Some	0.76 – 1.5
Moderate	1.51 – 3.00
Severe	≤3.00

Note: *dS/m at 25ºC = mmhos/cmLeaching required at higher range.Good drainage needed and sensitive plants may have difficulty at germination.

SALINITY HAZARD

The most influential water quality guideline on crop productivity is the water salinity hazard as measured by electrical conductivity (EC_w). The primary effect of high EC_w water on crop productivity is the inability of the plant to compete with ions in the soil solution for water (physiological drought). The higher the EC, the less water is available to plants, even though the soil may appear wet. Because plants can only transpire "pure" water, usable plant water in the soil solution decreases dramatically as EC increases. Actual yield reductions from irrigating with high EC water varies substantially. Factors influencing yield reductions include soil type, drainage, salt type, irrigation system and management.

Table. Potential Yield Reduction from Saline Water for Selected Irrigated Crops.

	% Yield Reduction			
Crop	**0 %**	**10%**	**25%**	**50%**
		EC_w		
Barley	5.3	6.7	8.7	12
Wheat	4.0	4.9	6.4	8.7
Sugarbeet	4.7	5.8	7.5	10
Alfalfa	1.3	2.2	3.6	5.9
Potato	1.1	1.7	2.5	3.9
Corn (grain)	1.1	1.7	2.5	3.9
Corn (silage)	1.2	2.1	3.5	5.7
Onion	0.8	1.2	1.8	2.9
Dry Beans	0.7	1.0	1.5	2.4

Note: EC_w = electrical conductivity of the irrigation water in dS/m at 25oC. Sensitive during germination. EC_w should not exceed 3 dS/m for garden beets and sugarbeets.

The amount of water transpired through a crop is directly related to yield; therefore, irrigation water with high EC_w reduces yield potential (Table). Beyond effects on the immediate crop is the long term impact of salt loading through the irrigation water. Water with an EC_w of only 1.15 dS/m contains approximately 2,000 pounds of salt for every acre foot of water. You can use conversion factors in Table to make this calculation for other water EC levels.

Other terms that laboratories and literature sources use to report salinity hazard are: salts, salinity, electrical conductivity (EC_w), or total dissolved solids (TDS). These terms are all comparable and all quantify the amount of dissolved "salts" (or ions, charged particles) in a water sample. However, TDS is a direct measurement of dissolved ions and EC is an indirect measurement of ions by an electrode.

Although people frequently confuse the term "salinity" with common table salt or sodium chloride (NaCl), EC measures salinity from all the ions dissolved in a sample. This includes negatively charged ions (*e.g.*, Cl, NO_3) and positively charged ions (*e.g.*, Ca, Na).

Another common source of confusion is the variety of unit systems used with EC_w. The preferred unit is deciSiemens per meter (dS/m), however millimhos per centimeter (mmho/cm) and micromhos per centimeter (µmho/cm) are still frequently used.

Table. Conversion Factors for Irrigation Water Quality Laboratory Reports.

Component	To Convert	Multiply By	To Obtain
Water nutrient or TDS	mg/L	1.0	ppm
Water salinity hazard	1 dS/m	1.0	1 mmho/cm
Water salinity hazard	1 mmho/cm	1,000	1 μmho/cm
Water salinity hazard	EC_w (dS/m) for EC <5 dS/m	640	TDS (mg/L)
Water salinity hazard	EC_w (dS/m) for EC >5 dS/m	800	TDS (mg/L)
Water NO_3N, SO_4-S,B applied	ppm	0.23	lb per acre inch of water
Irrigation water	acre inch	27,150	gallons of water

DEFINITIONS

Abbrev.	Meaning
mg/L	Milligrams per liter
meq/L	Milliequivalents per liter
ppm	Parts per million
dS/m	DeciSiemens per meter
μS/cm	MicroSiemens per centimeter
mmho/cm	Millimhos per centimeter
TDS	Total dissolved solids

SODIUM HAZARD

Infiltration/Permeability Problems

Although plant growth is primarily limited by the salinity (EC_w) level of the irrigation water, the application of water with a sodium imbalance can further reduce yield under certain soil texture conditions. Reductions in water infiltration can occur when irrigation water contains high sodium relative to the calcium and magnesium contents. This condition, termed "sodicity," results from excessive soil accumulation of sodium. Sodic water is not the same as saline water. Sodicity causes swelling and dispersion of soil clays, surface crusting and pore plugging. This degraded soil structure condition in turn obstructs infiltration and may increase run-off. Sodicity causes a decrease in the downward movement of water into and through the soil, and actively growing plants roots may not get adequate water, despite pooling of water on the soil surface after irrigation.

The most common measure to assess sodicity in water and soil is called the Sodium Adsorption Ratio (SAR). The SAR defines sodicity in terms of the relative concentration of sodium (Na) compared to the sum of calcium (Ca) and magnesium (Mg) ions in a sample. The SAR assesses the potential for infiltration problems due to a sodium imbalance in irrigation water. The SAR is mathematically written below, where Na, Ca and Mg are the concentrations of these ions in milliequivalents per liter (meq/L).

Concentrations of these ions in water samples are typically provided in milligrams per liter (mg/L). To convert Na, Ca, and Mg from mg/L to meq/L, you should divide the concentration by 22.9, 20, and 12.15 respectively. For most irrigation waters encountered in Colorado the standard SAR formula provided above is suitable to express the potential sodium hazard. However, for irrigation water with high bicarbonate (HCO_3) content, an "adjusted" SAR (SAR_{ADJ}) can be calculated.

In this case, the amount of calcium is adjusted for the water's alkalinity, is recommended in place of the standard SAR (see pH and Alkalinity section below). Your laboratory may calculate an adjusted SAR in situations where the HCO_3 is greater than 200 mg/L or pH is greater than 8.5.

$$SAR = \frac{Na^{+} meq / L}{\sqrt{\dfrac{\left(Ca^{++}{}_{meq/L}\right) + \left(Mg^{++}{}_{meq/L}\right)}{2}}}$$

meq/L = mg/L divided by atomic weight of ion divided by ionic charge (Na = 23.0 mg/meq, Ca = 20.0 mg/meq, Mg = 12.15 mg/meq) The potential soil infiltration and permeability problems created from applications of irrigation water with high "sodicity" cannot be adequately assessed on the basis of the SAR alone.

This is because the swelling potential of low salinity (EC_w) water is greater than high EC_w waters at the same sodium content. Therefore, a more accurate evaluation of the infiltration/permeability hazard requires using the electrical conductivity (EC_w) together with the SAR.

Table. Guidelines for Assessment of Sodium Hazard of Irrigation Water Based on SAR and EC_w.

	Potential for Water Infiltration Problem	
Irrigation water SAR	**Unlikely**	**Likely**
	——————EC_w (dS/m)——————	
0-3	>0.7	<0.2
3-6	>1.2	<0.4
6-12	>1.9	<0.5.
12-20	>2.9	<1.0
20-40	>5.0	<3.0

Many factors including soil texture, organic matter, cropping system, irrigation system and management affect how sodium in irrigation water affects soils. Soils most likely to show reduced infiltration and crusting from water with elevated SAR (greater than 6) are those containing more than 30 per cent expansive (smectite) clay. Soils containing more than 30 per cent clay include most soils in the clay loam, silty clay loam textural classes and finer and some sandy clay loams. In Colorado, smectite clays are common in areas with agricultural production.

Table. Susceptibility Ranges for Crops to Foliar Injury from Saline Sprinkler Water.

	Na or Cl Concentration (mg/L) Causing Foliar Injury			
Na concentration	<46	46-230	231-460	>460
Cl concentration	<175	175-350	351-700	>700
	Apricot	Pepper	Alfalfa	Sugarbeet
	Plum	Potato	Barley	Sunflower
	Tomato	Corn	Sorghum	

pH and Alkalinity

The acidity or basicity of irrigation water is expressed as pH (< 7.0 acidic; > 7.0 basic). The normal pH range for irrigation water is from 6.5 to 8.4. Abnormally low pH's are not common in Colorado, but may cause accelerated irrigation system corrosion where they occur. High pH's above 8.5 are often caused by high bicarbonate (HCO_3) and carbonate (CO_3) concentrations, known as alkalinity.

High carbonates cause calcium and magnesium ions to form insoluble minerals leaving sodium as the dominant ion in solution. As described in the sodium hazard section, this alkaline water could intensify the impact of high SAR water on sodic soil conditions. Excessive bicarbonate concentrates can also be problematic for drip or micro-spray irrigation systems when calcite or scale build up causes reduced flow rates through orifices or emitters. In these situations, correction by injecting sulfuric or other acidic materials into the system may be required.

Chloride

Chloride is a common ion in Colorado irrigation waters. Although chloride is essential to plants in very low amounts, it can cause toxicity to sensitive crops at high concentrations. Like sodium, high chloride concentrations cause more problems when applied with sprinkler irrigation. Leaf burn under sprinkler from both sodium and chloride can be reduced by night time irrigation or application on cool, cloudy days. Drop nozzles and drag hoses are also recommended when applying any saline irrigation water through a sprinkler system to avoid direct contact with leaf surfaces.

Table. Chloride Classification of Irrigation Water.

Chloride (ppm)	Effect on Crops
Below 70	Generally safe for all plants.
70-140	Sensitive plants show injury.
141-350	Moderately tolerant plants show injury.
Above 350	Can cause severe problems.

Boron

Boron is another element that is essential in low amounts, but toxic at higher concentrations. In fact, toxicity can occur on sensitive crops at concentrations less than 1.0 ppm. Colorado soils and irrigation waters contain enough B that additional B fertilizer is not required in most situations. Because B toxicity can occur at such low concentrations, an irrigation water analysis is advised for ground water before applying additional B to crops.

Table. Boron Sensitivity of Selected Colorado Plants (B Concentration, mg/ L*)

Sensitive	Moderately Sensitive	Moderately Tolerant	Tolerant
0.76-1.0	1.1-2.0	2.1-4.0	4.1-6.0
Wheat	Carrot	Lettuce	Alfalfa
Barley	Potato	Cabbage	Sugar beet
Sunflower	Cucumber	Corn	Tomato
Dry Bean		Oats	

Sulfate

The sulfate ion is a major contributor to salinity in many of Colorado irrigation waters. As with boron, sulfate in irrigation water has fertility benefits, and irrigation water in Colorado often has enough sulfate for maximum production for most crops. Exceptions are sandy fields with <1 per cent organic matter and <10 ppm SO-S in irrigation water.

Nitrogen

Nitrogen in irrigation water (N) is largely a fertility issue, and nitrate-nitrogen (NO_3-N) can be a significant N source in the South Platte, San Luis Valley, and parts of the Arkansas River basins. The nitrate ion often occurs at higher concentrations than ammonium in irrigation water. Waters high in N can cause quality problems in crops such as barley and sugar beets and excessive vegetative growth in some vegetables. However, these problems can usually be overcome by good fertilizer and irrigation management. Regardless of the crop, nitrate should be credited towards the fertilizer rate especially when the concentration exceeds 10 ppm NO_3-N (45 ppm NO_3^-).

6

Irrigation Water Requirement of Crops

Several experiments on various crops for their irrigation requirements have highlighted the following results.

WATER REQUIREMENT BY CROPS

The fundamental concepts of institutional economics may be said to be demand and supply. So it is reasonable to begin building the analytic framework of this book with farmers' demand for irrigation services. But who are these 'farmers'? Here it is worthwhile drawing a distinction between farming families and agribusinesses. With the former, a family is at the core of the labour process in crop production.

That farming family may or may not engage wage-paid employees, it may grow much of its own food as well as pursuing commercial farming and the size of its landholding is usually small. In contrast an agribusiness has no family at the core of the labour process, virtually all its staff are employees, little or no crop output is grown which is not sold and holdings are usually large. In this book, for convenience, I shall usually adopt the term farmer, referring to farming families and agribusinesses only where the specific context requires it. Similarly, 'farmer' will be applied in the case of share tenants who may be moved by their landlord from time to time from one plot to another.

We start with a community of farmers in a river basin, each of whom has a given hectarage of land with a known cropping pattern. In output terms the farmer seeks tonnage of the crop, product quality and crop composition and timing of the harvest appropriate to the demands of the market. On the demand side, a fundamental agronomic fact is that crop growth depends on water availability to each plant's rootzone. The soil water generated by precipitation may source this. But in those cases where soil water is known to be insufficient for crop needs between planting and harvesting, then irrigation is required. The same holds true where precipitation and soil water availability cannot be predicted with reasonable accuracy; in this case irrigation sources are necessary to make up for any deficiency that does occur, for example during a drought period. Where rainfall follows a seasonal cycle, as in the case study of the Lower Bhavani Project the demand for irrigation water is counter-cyclical.

Full irrigation is required when no crop can be reasonably grown unless water is supplied by human effort. It is necessary for virtually any productive agriculture in arid and semi-arid regions. *Supplementary* irrigation indicates a supply of water to crops to increase yields and to reduce the risks of crop failure where, under normal circumstances, reasonable yields can be expected. Crop water needs, whether met by precipitation or by irrigation, are of central interest to the agronomist.

These needs and those for irrigation water can be expressed by a cluster of terms. *Crop water requirements* can be defined as the depth of water needed to meet the evapotranspiration requirements of a disease-free crop, growing in large fields under non-restricting soil conditions and achieving full production potential under the given growing environment. Alongside this optimum volume we have *actual crop evapotranspiration*. Next comes *effective precipitation* defined as that part of total rainfall that can be beneficially used by crops. Lastly we have *net irrigation requirements,* the amount of irrigation water needed to supplement rainfall to meet crop water requirement, excluding irrigation water lost to nonbeneficial evapotranspiration and drainage.

Different crops have different crop water requirements. There the irrigated 'dry crops' were cotton, groundnuts and maize, whereas sugar cane was expected to consume five times as much water per hectare. When it is feasible, farmers tend to grow dry crops in the dry season and wet crops in the wet season. Actual evapotranspiration of a typical crop is about 500 mm (5,000 m^3/ha) between planting and harvesting.

In this book I shall repeatedly use various measures of irrigation efficiency and productivity, terms used interchangeably. The first of these, *E*1, is an output to input ratio known as *relative efficiency*. Hargreaves and Christiansen have shown that *E*1 is strikingly similar across crops. Their data derived from Davis, California (corn and alfalfa), Hawaii (sugar cane) and Delhi, India (alfalfa). Here the input is measured as actual water supplied divided by the physical optimum for plant growth and output is measured as actual crop tonnage divided by the physical maximum, with all other factors such as seed and fertilizer optimized.

*E*1 is dimensionless, a pure number. Between the values of about 15 per cent and 65 per cent of the water input variable, the relative efficiency function is virtually linear; here a ten-point increase in water supply is matched by about a ten-point gain in output. Beyond about 65 per cent, output gain from a one-point water supply increase begins to weaken and, by definition, it becomes negative beyond the *E*1 value of 1.0. Plants can be supplied with too much as well as too little water for their maximum growth. (Exceptionally, rice has the capacity to grow in saturated conditions.) Overwatering may take place as a result of uncertainties in irrigation distribution. In the reverse case, when water not land is the limiting input, farmers may spread their water over a larger crop area. How important, on a world scale, is irrigation in the supply of water for agriculture? van Hofwegen and Svendsen distinguish between four water management regimes: rain-fed, irrigation only, drainage only and irrigation and drainage.

They write: Of the 1,500 million hectares of global cropland some 250 million hectares (17 per cent) are irrigated and approximately 150 million hectares are provided with drainage infrastructure. Currently about 60 per cent of global food crop production originates from rainfed agriculture and the remaining 40 per cent from irrigated agriculture.

DETERMINANTS OF WATER EFFICIENCY

Water efficiency, in terms of these physical measures, has many determinants other than the volume of water applied at the rootzone. One of these concerns the timing of rainfall or plant irrigation. Crop water requirements vary across the season from planting to harvesting. For example, water stress at the flowering stage of maize reduces yields by 60 per cent, even if water is adequate during the rest of the crop season. Similarly, the need for water can be critical at seeding time for germination and when seedlings are transplanted to the field.

For these reasons cultivators need to maintain adequate control over irrigation supply so that, in the ideal situation, water inputs are available precisely when they are needed. The desire for precision in the timing, volume and physical placement of irrigation water has stimulated water-use technologies that promise to achieve this, particularly sprinkler, drip and trickle irrigation. In practice, without adequate management, the new technologies can be less efficient than simpler surface methods. Another determinant of water efficiency is the volume and quality of farmers' other inputs to their cultivation practices. Better seeds, the right quantity of the right fertilizer, biocides to reduce crop losses and plentiful skilled labour at the right time of the season all boost field and basin water efficiencies by boosting output volume.

Moreover, it has been the availability of irrigation services that has provided the historical platform for these other contributions to water efficiency. Irrigation reduced the risk that farmers faced in rain-fed agriculture with its intermittent drought and subsequent crop failures. This manufacture of risk *reduction* made the decision to spend on other inputs much more profitable. For example, the extraordinary advances in volume of output, labour productivity and sales in the Indian fertilizer industry from the mid-1960s was certainly underpinned by growth in the irrigated area.

$$E_2 = \left(\frac{N}{S_b}\right)$$

where E_2 is requirements efficiency, N is the net irrigation requirements and S_b is defined above. The difference between N and S_b is equal to the storage and distribution losses between source and field, non-beneficial evaporation at the field level and field drainage.

Note that E_1, *relative* efficiency, deals with a relative ratio between crop output and water use, whereas requirements efficiency deals with an *absolute* ratio between evapotranspiration and water supply. As with E_1, E_2 is also a pure number.

The ratio of N to S_b can be expressed in another way. Where water efficiency uses as its input measure the gross irrigation volume drawn off by rainfall collection, surface and groundwater abstraction, reused wastewater and reused drainage water, then the more of this volume that is *captured* for the crops' net irrigation needs the greater will be the measured water efficiency E_2.

This is of tremendous importance in increasing 'crop per drop'. It should be noted that the measurements required to estimate requirements efficiency are extremely demanding, including as they do crop water requirements, the irrigation water quantity needed to supplement rainfall at the plant's rootzone and the multifarious losses between the supply source of the irrigation flow and the rootzone.

Here, as in other ways, the nature of the knowledge developed and applied by professional economists and agronomists, for example and that o the farmer is likely to differ considerably. In my view the primary interest o the economist in understanding the irrigation and drainage sector should be the values, beliefs, motives and behavioural routines of the sector's actors amongst whom farmers are the protagonists. Only on this basis can an effective economics of irrigation be constructed.

THE COST OF WATER TO FARMERS

Physical measures of water efficiency can be of value to water resource planners when they draw up a scenario hydrosocial balance for the purpose of basin strategy formulation. For example, we can write as *tonnage efficiency*

$$E_3 = \left(\frac{T}{S_b}\right)$$

where T is the actual tonnage of crop output in a region in the baseline yea and S_b is the volume of irrigation water supplied at the regional level. In thi case estimates of future desired crop output could be divided by tonnag efficiency to derive a first estimate of required irrigation supply in agricultur in millions of cubic metres in the scenario year. But water efficienc measurement in this way may be of little account in planning by farming families or agribusinesses. Their behaviour is shaped above all by economic motives and therefore by cash costs of inputs and value of outputs, not by the ratio of tonnage to water quantity ratios of agronomy and resource planning. So it is necessary to consider what are *the cash costs to farmers of their use of water*.

The *first case* is the simplest. A public-sector irrigation authority provides water free (or virtually free) of cost. Malaysia is an example; here, a nominal charge is made for water in the annual fixed land tax payable by farmers. In 1997 this was 15 Malaysian ringitts, or about US$0.50, per month. In this limiting case the irrigation flows are treated by the farmer as a free resource, no more expensive than rainfall.

As an economic unit the farming family or agribusiness regards the water as simply a critical natural resource that, with precipitation, determines output. The *second case* occurs where the farmer provides his own supply of water through rainfall collection, appropriating surface water or pumping groundwater. But it gives nothing away to report that, for example, a family with its own tubewell must bear its installation and running costs. The family's cash outlays for its water supply are here very real, particularly if the price of energy is high.

In this example water can be measured not only in cubic metres but also in Indian rupees, Chinese renminbi, Pakistani rupees, US dollars, Mexican new pesos or what you will. Note that in this situation farmers also deploy their labour time in supplying themselves with water-the traditional Egyptian shadouf is a ready example. So there is a labour cost not expressed in cash but with a significant opportunity cost.

In this case the opportunity cost is the output value foregone or the leisure time sacrificed by the farmer as a result of committing time to supplying the farm with irrigation water. The *third case* in considering the cash costs of water to the farmer is where a volumetric price is charged by an irrigation authority. Here the uptake of water by the farm is measured in volume terms and a price per cubic metre (for example) is charged. (The water tariff per unit may vary with the total volume used and with the time of day or the season during which the water is supplied.) It is pricing of this kind, for goods and services in general, that is the basis of most academic microeconomic analysis in market societies.

The *fourth case* is where an irrigation authority, or its equivalent such as a water users' association, makes a charge on the farming family or agribusiness that correlates with its water consumption per hectare but is not in the form of a volumetric price as in the third case. In this fourth case the charge paid by the farmer varies with the volume of water used but in an *indirect* way. The Department also used a multiplier in calculating the final tax bill to irrigators of 355 Rs/ha for the presumed water requirements of 5,000

m^3/ha for cotton and 505 Rs/ha for the presumed water requirements of 12,500 m^3 /ha for paddy. Another example of the fourth case situation is where farmers pay cash for their supplies of irrigation water on the basis of the length of time for which they have access to that supply.

Julie Trottier has a case study of Jericho's Ein Sultan spring that illustrates this situation. If the flow of irrigation water per unit time did not vary, a time-allocation price and volumetric pricing would give the same outcome. But, of course, irrigation water flows *do* vary with time. Time allocation is also used by the Mexican water user associations. The best known of all time-allocation arrangements is the *warabundi* method of rotation described by Jeremy Berkoff for the North India Act systems.

Water is allocated in proportion to land and water deliveries are controlled by time; full flow in the watercourse being used by each farmer in turn. The *fifth and last case* is where an irrigation service fee is charged by the irrigation authority but the payment is for revenue-raising purposes alone and bears no relation, direct or indirect, to the volume of water used. From the point of view of water resource planners, in those cases where a cost-tool can influence farm use of water, it can be deployed to raise water efficiency. Where it contributes to an irrigation authority's revenue, it can be deployed to finance all or part of that authority's capital and operational costs. Only the third and fourth cases satisfy both criteria.

CROP SELECTION

TO OVERCOME SALINITY HAZARDS

Not all plants respond to salinity in a similar manner; some crops can produce acceptable yields at much higher soil salinity than others. This is because some crops are better able to make the needed osmotic adjustments, enabling them to extract more water from a saline soil. The ability of a crop to adjust to salinity is extremely useful.

In areas where a build-up of soil salinity cannot be controlled at an acceptable concentration for the crop being grown, an alternative crop can be selected that is both more tolerant of the expected soil salinity and able to produce economic yields. There is an 8-10 fold range in the salt tolerance of agricultural crops. This wide range in tolerance allows for greater use of moderately saline water, much of which was previously thought to be unusable.

It also greatly expands the acceptable range of water salinity (EC_w) considered suitable for irrigation. The relative salt tolerance of most agricultural crops is known well enough to give general salt tolerance guidelines. Figure presents the relationship between relative crop yield and irrigation water salinity with regard to the four crop salinity classes.

Table. Information Required on Effluent Supply and Quality

Information	Decision on irrigation management
Effluent supply	
The total amount of effluent that would be made available during the crop growing season.	Total area that could be irrigated.
Effluent available throughout the year.	Storage facility during non crop growing period either at the farm or near wastewater treatment plant, and possible use for aquaculture.
The rate of delivery of effluent either as m^3 per day or litres per second.	Area that could be irrigated at any given time, layout of fields and facilities and system of irrigation.
Type of delivery: continuous or intermittent, or on demand.	Layout of fields and facilities, irrigation system, and irrigation scheduling.
Mode of supply: supply at farm gate or effluent available in a storage reservoir to be pumped by the farmer.	The need to install pumps and pipes to transport effluent and irrigation system.
Effluent quality	
Total salt concentration and/or electrical conductivity of the effluent.	Selection of crops, irrigation method, leaching and other management practices.
Concentrations of cations, such as Ca^{++}, Mg^{++} and Na^{+}.	To assess sodium hazard and undertake appropriate measures.
Concentration of toxic ions, such as heavy metals, Boron and Cl^{-}.	To assess toxicities that are likely to be caused by these elements and take appropriate measures.
Concentration of trace elements (particularly those which are suspected of being phyto-toxic).	To assess trace toxicities and take appropriate measures.
Concentration of nutrients, particularly nitrate-N.	To adjust fertilizer levels, avoid over-fertilization and select crop.
Level of suspended sediments. clogging problems.	To select appropriate irrigation system and measures to prevent
Levels of intestinal nematodes and faecal coliforms.	To select appropriate crops and irrigation systems.

The following general conclusions can be drawn from these data:

- Full yield potential should be achievable with nearly all crops when using a water with salinity less than 0.7 dS/m,
- When using irrigation water of slight to moderate salinity (*i.e.* 0.7-3.0 dS/m), full yield potential is still possible but care must be taken to achieve the required leaching fraction in order to maintain soil salinity within the tolerance of the crops. Treated sewage effluent will normally fall within this group,

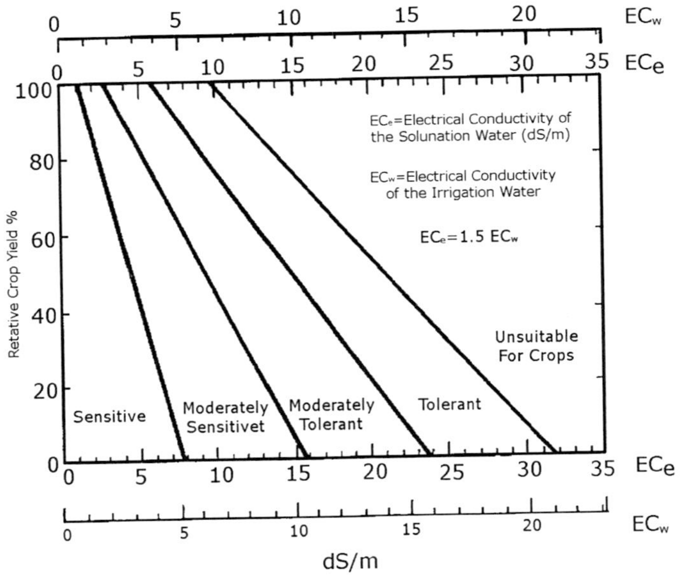

Fig. Divisions for Relative Salt Tolerance Ratings of Agricultural Crops

- For higher salinity water (more than 3.0 dS/m) and sensitive crops, increasing leaching to satisfy a leaching requirement greater than 0.25 to 0.30 might not be practicable because of the excessive amount of water required. In such a case, consideration must be given to changing to a more tolerant crop that will require less leaching, to control salts within crop tolerance levels. As water salinity (ECw) increases within the slight to moderate range, production of more sensitive crops may be restricted due to the inability to achieve the high leaching fraction needed, especially when grown on heavier more clayey soil types.
- If the salinity of the applied water exceeds 3.0 dS/m, the water might still be usable but its use may need to be restricted to more permeable soils and more salt-tolerant crops, where high leaching fractions are more easily achieved. This is being practised on a large scale in the Arabian Gulf States, where drip irrigation systems are widely used.

If the exact cropping patterns or rotations are not known for a new area the leaching requirement must be based on the least tolerant of the crops adapted to the area. In those instances, where soil salinity cannot be maintained within acceptable limits of preferred sensitive crops, changing to more tolerant crops will raise the area's production potential. If there is any doubt about the effect of wastewater salinity on crop production, a pilot study should be undertaken to demonstrate the feasibility of irrigation and the outlook for economic success.

TO OVERCOME TOXICITY HAZARDS

A toxicity problem is different from a salinity problem in that it occurs within the plant itself and is not caused by water shortage. Toxicity normally results when certain ions are taken up by plants with the soil water and accumulate in the leaves during water transpiration to such an extent that the plant is damaged.

The degree of damage depends upon time, concentration of toxic material, crop sensitivity and crop water use and, if damage is severe enough, crop yield is reduced. Common toxic ions in irrigation water are chloride, sodium, and boron, all of which will be contained in sewage. Damage can be caused by each individually or in combination. Not all crops are equally sensitive to these toxic ions.

However, toxicity symptoms can appear in almost any crop if concentrations of toxic materials are sufficiently high. Toxicity often accompanies or complicates a salinity or infiltration problem, although it may appear even when salinity is not a problem. The toxic ions of sodium and chloride can also be absorbed directly into the plant through the leaves when moistened during sprinkler irrigation.

This typically occurs during periods of high temperature and low humidity. Leaf absorption speeds up the rate of accumulation of a toxic ion and may be a primary source of the toxicity. However, urban wastewater may contain heavy metals at concentrations which will give rise to elevated levels in the soil and cause undesirable accumulations in plant tissue and crop growth reductions.

Heavy metals are readily fixed and accumulate in soils with repeated irrigation by such wastewaters and may either render them non-productive or the product unusable. Surveys of wastewater use have shown that more than 85 per cent of the applied heavy metals are likely to accumulate in the soil, most at the surface. The levels at which heavy metals accumulation in the soil is likely to have a deleterious effect on crops. Any wastewater use project should include monitoring of soil and plants for toxic materials.

WATER REQUIREMENT OF DIFFERENT CROPS

Climate, soil and water are the three basic resources which determine the nature of crops that can be grown successfully in a particular region. An efficient utilization of these resources is essential for optimum production of food and fibre for human life, feed for cattle and raw materials for industry. Climate determines the suitability of a region as a habitat for different flora and fauna, also the availability of water for production of Crops and other uses.

Under a given set of environmental conditions production of crops is limited by the availability of nutrients and water. Soil provides anchorage for the plants and serves as a reservoir of water and nutrient required by them.

Chemical fertilizer can supplement nutrients to poor soil, but there is no substitute of water for production of crops. As the resources are limited an efficient management is of utmost importance for sustaining and increasing agricultural production.

Competing demands of water for domestic use, sanitation, industrial and recreational purposes makes it more essential to maximize the efficiency of water for agricultural production. Agriculture consumes the largest amount of available water, yet, it uses water less efficiently. If food crisis is to be solved, there is no other alternative than to increase area under irrigation which can be made possible only when we use our present water resources most judiciously for irrigating our agricultural crops.

PADDY

Paddy is a semi-aquatic plant and covers about 35 per cent of irrigated area in the country. Different varieties have been evolved to suit different regions, season and water availability. Cultural practices like puddling and transplanting reduce percolation losses, weed growth but increase the availability of plant nutrients and regulate soil and water temperatures. It improves photosynthesis in the lower leaves due to reflected light from the water surface.

These operations may require about 200 -300 mm of water per hectare. Submergence below 50 mm for low land rice has been found to yield low. Recent researches have shown that continuous submergence throughout the growth period may not maximize the yield. Selective submergence during critical stages (initial tillering, panicle initiation to flowering) would be sufficient to maximize yield and to save water during the monsoon period. However, during summer, continuous submergence has to be followed for maximum yields, Proper drainage helps to remove the toxic substances and regulate the oxygen supply to the roots.

About 15 to 20 days prior to harvest, irrigations are stopped and water is drained to facilitate harvesting operations. A major problem in paddy irrigation is deep percolation losses which is 50 to 75 per cent of water applied. Reduction of deep percolation of water is generally achieved by soil manipulation of three types *viz.*

(i) Puddling (ii) compaction, and (iii) sub-surface placement of impermeable materials like bitumin and plastic films. Because of its prohibitive cost and mechanical difficulty, the third option is not practicable at farmers field. Farmers generally continue to irrigate rice till about 4 –7 days before harvest.

A recent study has stated that suspension of last irrigation for 14 – 17 days before harvest causes more uniform ripening of the crop and economises 16 cm of irrigation water. However, termination of irrigation 3 weeks before harvest may cause marginal yield decrease, but it will save about 20 cm of irrigation water. The critical stages for irrigation in rice crop are tillering, panicle initiation, flowering and grain filling stage. The irrigation requirement of paddy at different locations are given in table below.

Table. Water and Irrigation Requirement or Rice at Different Locations

Sl.No.	Place	Water requirement	Irrigation Requirement	Seasons
		(Mm)	(mm)	July/Aug/-Nov/Dec.
1	Kharagpur (WB)	1890	1440	Dec/Jan - Mar/April
		2150	N.A	
2	Cuttack (Orissa)	1300	790	June-September
		1190	780	Jan- April
3	Bhubaneshwar (Orissa)	1440	780	June-September
		1650	1630	Sept-Dec
4	Roorkee (U.P)	1620	750	June-October
5	Dhanauri (U.P.)	1630	910	June-October
6	Karnataka	1520	1170	June-October
7	Hyderabad (A.P)	N.A.	780	Mar-June
8	Coimbatore	N.A	1680	N.A. July/Aug Dec/January
9	Chalakudy (Kerala)	N.A	1520	Feb-May
10	Delhi	2400	1600	June-October
11	Ludhiana	N.A.	1240	June-October

Wheat

Wheat is the second most important crop of the country and the area under this crop has been on the increase during the last two decades. It is grown during winter and about 59 per cent of its area is irrigated. Recently introduced dwarf varieties are highly responsive to irrigation and fertilization. The critical stages for Irrigation are crown root initiation, tillering, jointing, flowering, milk stage and dough stage.

The crop develops a deep and dense root system unless restricted by an impeding layer in the profile. Therefore, it can effectively utilize profile-stored moisture, provided post -sowing irrigations are timed to encourage deeper proliferation of roots.

The fact is especially important for the major wheat zone of northern India, where at the time of its sowing the root zone profile is either charged deeply with moisture from the proceeding monsoon or with pre-sowing irrigation. Generally, 4-6 irrigations are found to yield in maximum productivity at about 40 to 50 per cent of depletion of available soil moisture. According to the availability of irrigation water, it may be scheduled as in table below.

Table. Number or Irrigations According to the Availability

Sl.No	One	Two	Three	Four	Five
1	CRI (21 DAS)	CRI (21 DAS)	CRI (21 DAS) Tillering (45 DAS)	CRI (21 DAS) Tillering (45 DAS)	CRI (21 DAS) Tillering (45 DAS)
2					Jointing (65 DAS)
3		Boot (75 DAS)	Boot (75 DAS)	Boot (75 DAS)	
4				(85 DAS)	Flowering
5			(100 DAS)	Milk (100 DAS)	Milk

Sorghum

Cultivation of sorghum is mostly confined to tropical and subtropical areas in Mabarashtra, Madhya Pradesh, Andbra Pradesh, Karnal Gujarat. Rajasthan, Uttar Pradesh, Tamil Nadu, Punjab and Haryana. Being the third important cereal of the country, the crop is planted in a total area of about 16 million hectares, of which 0.7 million hectares are irrigated. In Punjab, Haryana, Uttar Pradesh and some part of Rajasthan, it is mostly grown for forage rather than grain. The main growing season of the crop extends from March to October.

In southern states, it is also taken as a rabi crop during October to February. Experiments have shown that kharif crop do not require any irrigation, if there is sufficient rain during the season. It is a drought resistant crop and can withstand soil moisture depletion up to 75 per cent. Some. varieties of sorghum are of very short duration, adding to the adaptability, varied soil and water conditions. Seeding, flowering and sometimes milking are the critical stages when the crop is irrigated in case it is raised as irrigated crop.

Maize

Maize is grown both for grain and forage. Maize occupies an area of 5.9 million hectares. It is grown primarily as a kharif crop from March to October in Uttar Pradesh, Bihar, Madhya Pradesh, Punjab, Gujarat, Jammu and Kashmir, Himachal Pradesh, Orissa. Andhra Pradesh, Karnataka, Maharashtra and Haryana. Some also come under rabi crop in the southern, mid west states and more recently in Bihar, Uttar Pradesh and Punjab. Preliminary studies have shown that rabi crop has a higher judicious water management and is important for improving the productivity of both kharif and rabi crops.

The crop has early vegetative, tasselling and silking stages as critical periods. After dough stage, there is no need of irrigation. The permissible depletion of soil moisture may be 25 per cent in light soils and 50 per cent in sandy loam to loam soils. Some research findings have revealed that crop should not be subjected to excess water- soil submergence during early growth. If the draining out of water in case of heavy rains is delayed, 30 to 60 kg N/ha may be added immediately following drainage to retrieve the yield loss.

Bajra

Bajra is an important millet grown during kharif in warm areas with a seasonal rain of about 15-20 centimeters. It is cultivated on an area of 11.7 million hectares mostly on relatively light textured soils of Rajasthan, Haryana, Punjab, Uttar Pradesh, Kamataka, Gujarat and Tamil Nadu. In most of the areas it is sown with the onset of monsoon and no post-sowing irrigation is applied. However, the crop, particularly the hybrid strains, has been found to respond to irrigation.

Flowering and milking stages are the critical stages of irrigation for bajra crop. The crop gives the optimum yield with irrigation based on 75 per cent depletion of available soil moisture from the top 30 cm layer. Bajra is a drought resistant crop. If there is enough rain, kharif crop does not require irrigation. If there is no rain generally, two irrigations are required, first at flowering stage and second at the milking stage. If moisture is a limiting factor, irrigation should also be done at the time of ear head emergence because it is the most critical stage for moisture stress. Bajra does not tolerate water logging. So, do not allow rain water to remain in the field for more than a few hours. Proper arrangement for draining out of excess water must be made.

Barley

Barley is an important rabi cereal and is also used in brewery industry. It is grown on an area of only 1.75 million hectares in the states of Uttar Pradesh, Madhya Pradesh, Rajasthan, Bihar, Haryana, Punjab and Himachal Pradesh. Its cultivation is preferred to wheat under low fertility, low irrigation and late sown conditions.

A few irrigation studies based on growth stages of barley have shown that barley shows favourable response to irrigation particularly if the rainfall is low. Generally, it require two to three irrigations to give good yield. One extra irrigation shall be required to sandy soils.

If supply of water is inadequate, its efficiency should be increased by giving irrigation at critical stages of growth. If only one irrigation is available, it should be given near active tillering stage (30 -35 days after sowing). When two irrigations are available, one should be applied at active tillering and the other at flowering st3ge. On highly saline and sodic soils frequent light irrigation gives better result than few heavy irrigations.

Cotton

Cotton is cultivated primarily as a kharif crop on an area of 8 million hectares of which only 25 per cent is irrigated. Major cotton growing states are Maharashtra, Gujarat. Karnataka, Punjab, Madhya Pradesh, Andhra Pradesh, Rajasthan, Haryana and Tamil Nadu. The crop is sown in the hot months of April and May and is harvested during October to November in northwest and mid- west States and December to January in southern states. It requires more frequent irrigation in the southern states where the weather is warm and long season varieties are grown. In the Northwest. cotton generally needs 2 to 5 irrigations depending upon the soil type and amount and distribution of seasonal rains. In drier and ill-distributed rainfall conditions, the crop needs 6 to 8 irrigations with 50 to 75 per cent soil moisture depletion.

Delayed irrigation prevents the plants from making excessive vegetative growth. The first irrigation should be given 40-45 days after sowing and subsequent irrigations should be light and be given at an interval of two to three weeks. The crop should not be allowed to suffer from water stress during flowering and fruiting period, other wise excessive shedding of flower buds and young bolls may occur resulting in the loss of yield. Cotton during its early growth is very sensitive to water stagnation for long periods. Therefore, proper drainage should be done.

Sugarcane

Sugarcane is an important crop of tropical areas. In India, it is grown on an area of about 3.19 m. ha. in Uttar Pradesh, Karnataka, Haryana. Bihar, Punjab, Gujarat and Madhya Pradesh. It is planted during January to March and is harvested after 10 –18 months in different parts of the country. It has a fibrous root system which can penetrate as deep as 2.5 m in well–drained soils devoid of any restricting layer and salinity. Thus, the crop utilizes most of the moisture stored in the root -zone. In the major sugarcane growing areas, 2–3 months dry and hot pre-monsoon period of the growing season is very crucial from the irrigation point of view. During this period the young crop is adversely affected by severe soil moisture deficiency and excessively high soil temperature. In the rainy period, there is little need for irrigation unless the rains are too scanty and erratic. During post– monsoon growth period crop again needs irrigation adequately to meet its evapo-transpiration needs and in some cases to mitigate damage by frost.

The life cycle of sugarcane plant is divided into four distinct phases namely germination phase (from planting to 60th day); formative phase (from 60th to 130th day); grand growth phase (from 130th to 250th day) and maturity phase (250th to 365th day). The water requirement during formative phase and grand growth phase is maximum. Generally under north Indian condition, the water requirement of grand growth phase is met with rain, while the requirement of formative phase (during pre- monsoon period) has to be met through irrigation. The total water requirement for sugarcane varies from 200–300 centimeters. In northern India, autumn cane requires on an average 7 irrigations, 5 before monsoon and 2 after monsoon. Spring planted crops may be irrigated only six times.

In drier climate and light soils, the crop may require 8 to 10 irrigations. In each irrigation, 3 acre inch of water should be applied. In Tarai areas, 2 to 3 irrigations before and one irrigation after monsoon are sufficient. Drainage is also equally important in waterlogged areas. Drain excess water from the sugarcane field if they are flooded during the rainy season. Due to waterlogged condition, the quality of cane deteriorates greatly. Drainage greatly helps in increasing the yield and sucrose content of the cane.

Groundnut

Groundnut is the most important oilseed crop of the country. It is mainly grown in Gujarat, Andhra Pradesh. Tamil Nadu, Maharashtra, Karnataka, Madhya Pradesh, Uttar Pradesh, Orissa, Rajasthan and Punjab. Raised as kharif crop in warm areas of relatively high rainfall, it often suffers from periodic water deficits during long rainless intervals. This occasional moisture deficiency is one of the important factors contributing to low yield of kharif groundnut. Consequently provision of irrigation to groundnut would greatly aid in improving its yield. Even now the farmers of southern states are growing groundnut during rabi as an irrigated crop.

Being a rainy season crop, groundnut does not require irrigation. However, if dry spell occurs, irrigation may become necessary. One irrigation should be given at pod development stage. In the southern part of the country where groundnut is grown as rabi crop, 3–4 irrigations are necessary. Give the first irrigation at the start of flowering and subsequent irrigations whenever required during the fruiting period to encourage peg penetration and pod development. The last irrigation before harvesting will facilitate the full recovery of pods from the soil.

Mustard and Rapeseed

India is a major producer of mustard and rapeseed crops. These crops are primarily raised as rainfed crops during the rabi season on an area of 4.38 million hectares in Uttar Pradesh, Rajasthan, Madhya Pradesh, Assam, Haryana, West Bengal, Gujarat, Orissa, Punjab and BihaI. Due to low and uncertain rainfall during their growing season, these crops generally show favourable response to irrigations.

It has been told that these crops need 1–4 irrigations, depending upon the soil moisture storage in the profile and the prevailing weather. Pre-bloom and pod filling stages are considered to be critical stages, therefore, irrigations at these stages are beneficial.

Gram

Gram is the most important pulse crop which is grown on an area of 3.5 million hectares mostly as a rainfed rabi crop in Madhya Pradesh, Rajasthan, Uttar Pradesh, Haryana, Maharashtra, Punjab, Karnataka etc. The crop has a deep tap root system and can make an efficient use of the profile–stored water in the well–drained alluvial soils of northern states and retentive clay or clay loam soils of mid–west and southern states. Although, water requirements of the crop are relatively low, it often suffers from periodic water deficits owing to low and uncertain rains in the rabi season. Consequently, the average yield of the rainfed crop is low.

With recent advancement in irrigation facilities, a part of rainfed area under gram can be brought under irrigation for improving its yield. Therefore, there is need to conduct studies to determine irrigation requirements of the crop in various agro– climatic region. If irrigation facilities are easily available, give a pre–sowing irrigation which will ensure proper germination and smooth crop growth. If winter rains fail, give one irrigation at pre–flowering stage and one at pod development stage. In no case first irrigation should be given earlier than four weeks after sowing. No irrigation should be given at flowering time. A light irrigation should always be given because heavy ones are harmful. Excess of irrigation enhances vegetative growth and depresses the yield.

CONDITIONS OF IRRIGATION WASTEWATER

Irrigation may be defined as the application of water to soil for the purpose of supplying the moisture essential for plant growth. Irrigation plays a vital role in increasing crop yields and stabilising production. In arid and semi-arid regions, irrigation is essential for economically viable agriculture, while in semi-humid and humid areas, it is often required on a supplementary basis.

At the farm level, the following basic conditions should be met to make irrigated farming a success:

- The required amount of water should be applied;
- The water should be of acceptable quality;
- Water application should be properly scheduled;
- Appropriate irrigation methods should be used;
- Salt accumulation in the root zone should be prevented by means of leaching;
- The rise of water table should be controlled by means of appropriate drainage;
- Plant nutrients should be managed in an optimal way.

The above requirements are equally applicable when the source of irrigation water is treated wastewater. Nutrients in municipal wastewater and treated effluents are a particular advantage of these sources over conventional irrigation water sources and supplemental fertilizers are sometimes not necessary. However, additional environmental and health requirements must be taken into account when treated wastewater is the source of irrigation water.

AMOUNT OF WATER TO BE APPLIED

It is well known that more than 99 per cent of the water absorbed by plants is lost by transpiration and evaporation from the plant surface. Thus, for all practical purposes, the water requirement of crops is equal to the evapotranspiration requirement, ET_c. Crop evapotranspiration is mainly determined by climatic factors and hence can be estimated with reasonable accuracy using meteorological data. An extensive review of this subject and guidelines for estimating ET_c, prepared by Doorenbos and Pruitt, are given in Irrigation and Drainage Paper 24.

A computer programme, called CROPWAT, is available in FAO to determine the water requirements of crops from climatic data. The water requirements of some selected crops, reported by Doorenbos and Kassam. It should be kept in mind that the actual amount of irrigation water to be applied will have to be adjusted for effective rainfall, leaching requirement, application losses and other factors.

Quality of Water to be Applied

Irrigation water quality requirements from the point of view of crop production. The guidelines presented are indicative in nature and will have to be adjusted depending on the local climate, soil conditions and other factors. In addition, farm practices, such as the type of crop to be grown, irrigation method, and agronomic practices, will determine to a great extent the quality suitability of irrigation water. Some of the important farm practices aimed at optimising crop production when treated sewage effluent is used as irrigation water will be discussed in this chapter.

Scheduling of Irrigation

To obtain maximum yields, water should be applied to crops before the soil moisture potential reaches a level at which the evapotranspiration rate is likely to be reduced below its potential. The relationship of actual and maximum yields to actual and potential evapotranspiration is illustrated in the following equation:

$$\left(1-\frac{Y_a}{Y_m}\right)=\left(1-\frac{ET_a}{ET_m}\right)$$

where:

Y_a = actual harvested yield

Y_m = maximum harvested yield

ky = yield response factor

ET_a = actual evapotranspiration

ET_m = maximum evapotranspiration

Several methods are available to determine optimum irrigation scheduling. The factors that determine irrigation scheduling are: available water holding capacity of the soils, depth of root zone, evapotranspiration rate, amount of water to be applied per irrigation, irrigation method and drainage conditions.

IRRIGATION WATER USE AT KEVERAL FARM

Keveral Farm is located close to Seaton on the low hills rising from the south coast of Cornwall in the UK. The farm is worked and managed by some seven or eight persons and is legally constituted as a co-operative company limited by guarantee known as Keveral Farmers Ltd. This group of co-operators sits neatly between the categories of farming family and agribusiness.

The farm has 30 acres (12.2 ha), of which just 5 acres (2.0 ha) are irrigated. But the number of crops per year varies from one to five depending on the product, so that the gross irrigated acreage is about 12 acres (4.9 ha). Gross irrigation area is equal to the net irrigation area (*i.e.* the actual irrigation area) multiplied by the average number of times that crops are grown per year on it. An actual irrigation area of 1 acre (0.4 ha) would be counted as a gross irrigation area of 2 acres (0.81 ha) if it were cropped twice in the year. Keveral Farmers Ltd mainly produces vegetables and soft fruits: beetroot, carrots, leeks, onions, parsnips, potatoes, swedes, turnips, broccoli, cabbage, cauliflower, kale, oriental brassicas, spinach, asparagus, aubergine, beans, broadbeans, courgettes, cucumber, custard marrow, herbs, lettuce, mushrooms, peas, peppers, tomatoes, strawberries and raspberries. All the produce is grown organically and the farm is managed following the aims and principles of the Soil Association, which certificates Keveral as qualifying as organic under the Association's demanding criteria. All the crops grown are irrigated.

About 3 per cent of output meets the personal needs of the co-operators; some goes to cafés and market stalls; and about 90 per cent is sold through the 'organic veggie boxes' scheme. Under this arrangement about 180 boxes of fresh, organic produce from the farm is packed each week and delivered by the co-operative to meet the orders of individual households either at their homes or at common drop-off points, such as the Environment Agency's offices in Bodmin. These deliveries are made to the villages and small towns close to the farm such as Looe, Downderry, Sheviock and Morval. Product quality is critical-in terms of freshness and organic sourcing-and purchasers are aware that this is rarely compatible with the cosmetic appearance of supermarket sales of agrochemically based outputs. Boxes come in three prices:

£5, £7 and £9. The farm has to meet the transport costs of their distribution network but receives 100 per cent of the retail price paid by the customer. The timing of crop availability is also critical-in this case to maintaining the customer base. Intermittent failure to deliver the weekly veggie boxes would quickly lead to loss of sales and termination of the farm's income stream. Admittedly there is a 'hunger gap' each year between late April and early June when little produce comes off the fields. Over the course of the year a substantial proportion of the boxes' contents is made up from other organic sources in Cornwall, Devon and as far away as Egypt and Morocco. Product variety is an important marketing advantage and this accounts for the wide assortment of vegetables grown.

Keveral Farm has four water sources. *The first* is a leat, the Cornish term for a contour ditch and this drains the hillside marking the farm's south- and west-aligned boundaries. Many decades ago an overshot water wheel turned here. The leat drains several small ponds and finally its discharge seeps down to the River Seaton below. The farm produces its own seedlings and they may be started off in the main pond. Otherwise the leat plays no part in the irrigation cycle.

The *second* supply source is a 43-m borehole driven into the sandstone aquifer. This supply point is used for irrigation only in emergencies. The *third* flow is captured by harvesting the rain that falls on the farmhouse. Gutters lead the precipitation to a downpipe that slopes directly into the 'new' storage tank. The roof captures 150 m^3 in an average year, when rainfall in this part of Cornwall is some 900 mm. The *fourth* water source comes from a stream that drains the north boundary of the farm. Part of its flow is diverted into three concrete tanks and from here a diesel-engine-powered pump raises the flow to the 'old' storage tank at the highest point on the farm, close to the farmhouse. The third and fourth flows dominate the supply side of Keveral Farm's irrigation cycle. Distribution from the two storage tanks to the crops is entirely by gravity.

The co-operator who was my principal informant during the fieldwork in May 2000 did not know how much water in total was supplied to the farm nor, specifically, how much water was supplied for irrigation purposes. None of the four flows referred to above is metered. There is a parallel here with farming families and agribusinesses in the Curu Valley in the previous case study. However, knowing the capacity of the two storage tanks (10m^3 and 8 m^3) and estimating how many times per week these tanks are emptied for irrigation purposes at different times of the year, my informant calculated an annual rate of supply from storage to the crops of about 600 m^3: 500 m^3 drawn from the stream and 100 m^3 from rainwater collection. Irrigation at Keveral Farm takes two forms: on open fields and in five polythene greenhouses, one of which is twin-span. The use of irrigation water on the former compensates for periods of dry weather. Irrigation gets under way after any spell of 7-9 days without rain. As indicated above, average precipitation in the area is plentiful and so the irrigation requirements of field crops are modest. Swales are used to reduce surface run-off and soil erosion. The greenhouses act as a means of climate control at the microscale. Because no rain enters them, crop water requirement is met entirely by irrigation. Given the farm's location, ambient control is a prerequisite to produce the volume, quality and variety of produce that the marketing of organic vegetables requires. The greenhouses limit the temperature range to which the plants are exposed, thereby eliminating this component of farm risk.

They create a warm and humid environment that is particularly suitable for the 'Mediterranean' crops; it is also a microclimate that checks the depredations of pests such as the red spider mite. The greenhouses shelter the growing plants from the often-blustery winds across the coastal hills-winds that can flatten vegetables or inhibit their growth. Finally, greenhouses keep the rain off! This discourages the entry of hungry legions of snails and slugs, it prevents plants receiving excess water that reduces output quality and it gives the co-operators a small warm world of work when lashing rain makes open-air labour unpleasant and inefficient.

My informant does not know the crop water requirements of any of the farm's produce, at least not in the agronomic sense of being able to cite annual requirement by crop per year in cubic metres per acre. In another, non-statistical sense, he understands the crops' water needs very well-through experience, trial and error and day-to-day observation, such as by thrusting his fingers into the soil to check how damp or dry are the top 2-3 inches (5-7 cm). For example, a crop of tomatoes was receiving irrigation water for 1 h in 24, but yellowing of the leaves due to overwatering was taking place so the irrigation schedule was changed.

Too much watering in the greenhouses in the winter months triggers fungal disease. We calculated the actual volume of water use as follows Outflow from the two main tanks for irrigation was some 600 m^3 in 1999 2000. Losses between the tanks and the fields or greenhouses is small-about 5 per cent. Leaching requirements are zero in organic farming. About 10 per cent of the gross supply probably represents the overwatering associated with introduction of new systems and insufficient staff training. This gives about 540 m^3/year for crop water requirements, about 570 m^3/year in actual evapotranspiration and about 48 m^3/ acre/year of water used in terms of the gross acreage.

A ratio of crop tonnage to water used has no practical meaning at Keveral Farm. Output is calculated in a way that relates to marketing, in weight (or units) per week per crop, such as 300 lbs (135 kg) of carrots/week or 25 lettuces/week. Planning and planting is done on the same basis. I now consider the cost of irrigation water to Keveral Farmers Ltd. The *capital* costs of supply are the rainfall collection infrastructure, the diesel engine and pump at th stream, a spare pump, the two storage tanks and the plastic-piped water distribution network to the fields. The *current* costs of own-supply in this case are the co-operators' time in maintenance of the supply network, spare parts and diesel fuel and lubrication oil for the diesel engine.

The only operating costs requiring money payment that have a volumetric relation to the quantity of water appropriated is that for diesel fuel, lubrication oil and spare parts, particularly for the pump. The co-operators did not know what this amounted to for pumping their water from the stream. The calculation was made for my benefit and the figure for this marginal cost is £0.22 (US$0.36) per cubic metre of water supplied to the 'old' storage tank at the top of the farm. The fact that the statistic was not known and is not used at Keveral suggests that the elasticity of demand for water is negligible. As a rule of thumb in the economics of irrigation, one can assume that, where a farming family or agribusiness knows neither the price nor the marginal cost of its water supply, the elasticity of demand is zero.

The expenditure of the co-operators' own labour-time in the operation and maintenance of their irrigation water supply was clearly important to them. However, irrigation also brings with it labour-time reduction in the cultivation process. This is because, for regularly irrigated crops on this farm, the distribution of organic fertilizer to the plants is carried out by attaching a bottle or bag containing the fertilizer at the head of the on-field irrigation network. As a result, the labour costs of fertilizer application are virtually eliminated. Farm budget recalculated from Keveral Farm's own cash-flow accounts as shown in Table. Total costs are some £62,000, of which 86 per cent is made up by vegetable purchases, wages, taxation and capital repayments. Total income is some £82,000, of which 91 per cent is made up of vegetable re-sales, miscellaneous products and irrigated crop sales. The purchase and re-sale of vegetables bought in from other organic farms is the largest single item for both the expenditure and income columns.

These vegetables plus the Farm's irrigated output are, of course, the source of the organic veggie boxes. Clearly, Keveral Farmers Ltd could raise its output in value-added terms through substituting these purchased inputs by additional output of its own produce. Gross margin per acre can be calculated as total income (excluding grants) minus prime costs divided by the farm's 30 acres (12.2 ha). The figure is £810. The gross margin for the 5 acres (2.0 ha) of irrigated land could not be calculated because prime costs are not separable into irrigation/ non-irrigation outputs. We can now also calculate one of the most widely used efficiency measures in irrigation economics. I shall call this E_7, *turnover efficiency*.

$$E_7 = \frac{I}{S}$$

where I is gross sales come (turnover) received by the farm for its irrigated produce. In this case the value is £19,331/600 m^3 or £32/m^3.

OPTIMISING IRRIGATION FOR AGRICULTURAL WATER MANAGEMENT

The basic principles of irrigation are quite simple. However, the practical application of these principles can provide complications. Crops are exposed to energy originating from the sun, and/or wind. Much of the energy is dissipated by evaporating water from leaf surfaces, which is commonly referred to as transpiration. Water may also be evaporated from the soil surface and the combination of evaporation from leaf and soil surfaces is evapotranspiration (ET).

Water transpired from the leaf surface is replaced by extracting water from the soil and transporting it through the root and stem system. The soil serves as a reservoir from which the plant extracts water to meet the demands of transpiration. The quantitative storage capacity depends on both the type of soil and the rooting depth of the crop plants. Soil texture and structure which result in large pore sizes, have lower storage capacity than soils with smaller pore sizes. More deeply rooted crops have a higher storage capacity as compared to shallow rooted crops. Irrigation is the practice of recharging the storage capacity that has been depleted by ET when natural precipitation is not adequate to meet the ET demands

The time to irrigate is usually before the soil is too dry for water to be extracted by the plant at a rate to meet the transpiration rate. If the rate of water movement to the leaf surface is less than the rate of transpiration, the plant responds by closing stomata to reduce the water loss. Carbon dioxide (CO_2) used for plant photosynthesis enters the plant through these same stomata. Therefore, closure of the stomata reduces water loss but also reduces CO_2 intake and therefore reduces the rate of photosynthesis. Thus the plant has a dual mechanism for protecting itself under a limited water supply. It reduces transpiration, thus attempting to maintain turgidity. At the same time it reduces photosynthesis, which would otherwise increase plant surface area causing greater interception of energy and more transpiration.

Numerous studies have reported that total dry matter production in plants is linearly related to ET. This observed result is consistent with the mechanism that the plant uses to protect itself against inadequate water. However, the part of some crops that is sold is not linearly related to total dry matter production. For these crops the marketed product might be achieved or possibly increased by allowing the soil to dry to a level reducing transpiration at specific time periods. Dr. Goldhamer will present the application of this concept.

The combination of a soil and crop rooting system that has a low storage capacity dictates the use of smaller quantity, more frequent irrigations; whereas a larger storage capacity allows less frequent though higher quantity irrigations. All irrigation waters contain some dissolved salts. Plants transpire pure water causing the soil solution to become concentrated with salts as transpiration proceeds. Because of this effect, occasionally irrigation must not only recharge the storage capacity, but additional water may be necessary to leach excessive salts from the root zone. The concentration of salts in the irrigation water and the crop tolerance to salinity dictate the amount of leaching. Irrigation management entails consideration of salinity as well as water content in the root zone.

7

Waterlogging System

Most anti-dam movements in india have emerged from move ments of people facing displacement due to the submergence of large areas upstream of dam sites. These struggles are expressions of a conflict of interests between those who bear the social and ecological costs of dams and those who benefit from them. However, large dams have diverse and complex ecological impacts. and they often generate environmental costs for those very groups who are supposed to be the beneficiaries. Waterlogging and salini sation are twin problems caused by the wasteful use of water.

Waterlogging is caused by the interaction of a large number of factors such as irrigation intensity, soil characteristics, drainage. seepage from reservoirs, distributaries and field channels. Since large-scale irrigation systems are linked to the uniformity of water distribution, which enforces the uniformity of cropping patterns. and uniformity in the landscape waterlogging becomes inevitable in areas with undulating topography and water retentive soils

. In such cases, farmers who are supposed to be the 'beneficiaries' become victims of irrigation projects and irrigation authorities. In areas where irrigation has led to the transformation of productive lands into waterlogged wastelands, conflicts arise between farmers and the state. The 'Mitt) Bachao Andolan' in the Tawacomman area is an example of such conflicts." In the Krishna basin, conflic generated by irrigation projects were highlighted by the farmer agitations in the command area of the Malaprabha project. The Malaprabha project was completed in 1972-73.

With the introduction of canal irrigation, nearly 2,364 hectares of land the project area has become waterlogged and saline' Before the introduction of perennial irrigation, the undulating semi-arid land in the project area wa used for growing water prudent crops like jowar and pulses. Due to a sudde change from rainfed agriculture to intensive canal irrigation, the low lyin areas have become waterlogged. The cultivation of water demanding crop like hybrid cotton has aggravated the problem. In addition, seepage fro canals has also raised the water level.

The Malaprabha project includes a storage dam of 1,068 million cum capacity near Saundatti in Belgaum district which feeds the 138 km Malaprabha right bank, 168 km left bank canal, and the Kolachi right bank canal. The soils in the command area are black cotton soils, which have high water retention capacity and are prone to waterloggin, Intensive irrigation of black cotton soils has been known to be prescription for creating wastelands. While irrigation has bet viewed as a means to improve land productivity, in cases like if Malaprabha command area, it has led to the destruction of productivity.

Further, the shift from rainfed food crops to an irrigated cash crop like cotton was expected to improve the prosperity of farmer However, it led to indebtedness as well as loss of fertile fan. through waterlogging. To utilise the irrigation waters, farmers began to cultivate 'Varalaxmi' cotton which was initially sold at ₹ 1,000 per quintal. Farmers took loans from banks to develop land, purchase seeds, chemical fertilisers and pesticides. The total loan taken by the farmers increased from ₹ 50 lakhs in 1974 to over ₹ 5.5 crores by 1980. The prices of chemical fertilisers increased from ₹ 75 to ₹ 103 per bag. The cost of the Varalaxmi seed increased from ₹ 60 to ₹ 170 per kg. In the meantime the price of cotton crashed from ₹ 1,000 per quintal in 1974 to ₹ 350 per quintal in 1980.

While farmers were caught in the trap of unfulfilled commercial promise, banks demanded repayments of loans, and the irrigation authorities demanded a development tax known as betterment levy of ₹ 500 to ₹ 1,600 per acre. The water tax was raised from ₹ 18 to ₹ 30 per acre for jowar, ₹ 18 to ₹ 50 per acre for Varalaxmi. A tax of ₹ 10 per acre was fixed even if water was not utilised. For the farmers, this amounted to gross injustice, since they had not benefited from the irrigation project. In addition, the compensation for acquiring land for the dam and canals had not been paid to 75 per cent of the farmers even after seven to eight years.

The farmers therefore organised themselves as the Malaprabha Niravari Pradesh Ryota Samvya Samithi' (Co-ordination Committee of Farmers of Malaprabha Ittihsyrf Area) in March 1980. When the local authorities did not pay heed to the farmers' demands, they launched a non-cooperation movement for nonpayment of taxes. The authorities responded by refusing to issue the certificates required by the farmers' children in order to in schools and colleges. On 19 June, the farmers went on a hunger strike in front of the Tebsildar's office in Naragund town. On 30 June, 10,()00 farmers collected to support those on hunger strike.

On 7 July, a massive rally was organised in Navalgund, and the farmers went on a hunger strike. Seeing that no response was forthcoming from the authorities, the farmers organised a 'bunch' on 21 July. When 5,000 to 6,000 farmers had gathered in Navalgund, their tractors were damaged and the rally was stoned. The protest then took a violent turn. The angry farmers seized the irrigation office department, burnt down one truck and fifteen jeeps. The police in turn opened fire and a young boy, Basappa Shivappa of Algavadi, was killed on the spot. In Naragund town, the police opened fire at a procession of 10,000 people, shooting one youth. The protesting farmers responded by beating a police officer and a constable to death.

The protests rapidly spread to Ghataprabha, Tungabhadra, and other parts of Karnataka. During the protests thousands of farmers were arrested and forty were killed. Finally, the government had to put a moratorium on the collection of water taxes.and the betterment levy. According to a rough estimate, the concessions granted to farmers to end the Malaprabha agitation amounted to ₹ 85 crores.

However, the high costs of irrigation in the Malaprabha project have been forgotten. No lessons have been drawn for planning water projects, and bureaucrats, technocrats and politicians continue-to get carried away by the euphoria for large dams and intensive irrigation projects. The creation of waterlogged wasteland through intensive irrigation is not specific to the Malaprabha command area. Compared to other projects in the Krishna basin, waterlogging, salinity and alkalinity are most serious in the Tungabhadra project.

Nearly 1,500 hectares of land is likely to become waterlogged under the left bank canal. In the right bank canal 6,000 hectares have been affected by waterlogging. Under the right bank high level canal 12,000 hectares have been affected by waterlogging. In all, 19,500 hectares have been destroyed by waterlogging in the Tungabhadra project within an irrigation period of thirty-five years.

In the Bhadra project, of 1,24,392 hectares irrigated, 7,900 hectares have been devastated by waterlogging. In the Malaprabha project, of the total potential of 2,12,086 hectares, only 12,186 hectares have been actually irrigated. Of this irrigated area, 50 per cent has been waterlogged. In the Ghataprabha project, where a higher average has been irrigated than what was actually planned, out of 3,58,542 hectares irrigated, 19,948 hectares have been devastated by waterlogging. In Andhra Pradesh, under the Nagarjunasagar project (NSP) the groundwater level has risen alarmingly within ten years, thereby indicating a trend towards waterlogging. In Maharashtra, the Maharashtra Irrigation Commission claims that 28,000 hectares of land have been affected by waterlogging in the Deccan Canals, *i.e.*, Nira and Mutha Canals.

The irrigation commission of 1976 had estimated the total waterlogged area in the basin at 7,828 hectares (6,583 hectares in Karnataka and 1,245 hectares in Maharashtra) and 15,502 hectares affected by salinity. At present 4,45,985 hectares have been affected by salinity. Waterlogging is ecologically linked to large dams because large darns involve the transport of huge quantities of water for intensive irrigation. In fact, the primary rationale given in defence of large dams is to induce a shift from protective irrigation, which is ensured by indigenous irrigation systems, to intensive irrigation for commercial crops. The inevitable ecological impact of overuse of water for irrigation is a build up of water beyond the drainage capacity of the ecosystem. The need for artificial drainage systems arises because the natural drainage processes of the local ecosystem are violated.

Waterlogging is thus a symptom of the conflict between water use in the commerciaVmarket economy, and water use for the maintenance of the water cycle including a balance between water entering an ecosystem and water leaving it. By violating the ecological laws of water flow, large dams lead to ecological destruction on the one hand and political conflict on the other.

EFFECTS OF WATERLOGGING

Water logging leads to salinity. When the water table rises up or if the plant roots happen to come within the capillary fringe, water is evaporated through capillarity. Thus, with the upward flow of water from the water table to the land surface during evaporation, the dissolved salts present in the water are carried to the surface resulting in deposition of salts in the root zone of crops, which eventually reduces the osmotic activity of the plants leaving the plants to fade away.

Thus water logging and salinity may be treated as twin problem, which deserve special treatments based on the local conditions and soil texture, structure and topography.

Remedial Measures Needed

- Lining of canal system
- Conjunctive use of surface and ground water
- Improvement of drainage system
- Improvement in water management practices
- Development of command area
- Bio-drainage system
- Research and development programme.

WUAs and Their Progress

The Command Area Development Programme started in 1974 envisaged the participation of farmer organisations as a necessary step to run the micro system.

The Committee on Pricing of Irrigation Water (1992) also recommended farmers participation in the management of irrigation systems. Inspite of the growing realisation of the urgent need for farmers' participation in the management of irrigation, the progress has been slow so far. It is estimated that today only 8,04,000 hectares are being managed by WUAs.

Table. Water User Associations

States	No of WUAS	Area covered ('000 ha)
Andhra Pradesh	10292	4800.0
Assam	509	24.09
Bihar	32	110.50
Chattishgarh	945	1119.0
Gujarat	576	96.68
Karnataka	2318	1528.27
Kerala	4079	149.09
Madhya Pradesh	10282	1502.0
Maharashtra	1329	490.0
Orissa	13434	1054.0
Rajasthan	506	619.65
Punjab	957	116.95
Tamil Nadu	1566	599.40
Uttar Pradesh	279	10.55
West Bengal	10000*	37.0*

* Under MI and RIDF schemes;

The major reasons for the tardy progress in the implementation of the PIM are: farmers dependent on the Government and farmers are reluctant to adopt participatory approach unless deliveries of water can be made flexible, practical and responsive to the need. It is believed that only when the operation and maintenance activities are fully handed over to the farmer's organisation, it will create a feeling in the minds of the farmers, that the water resource system is no more a Government property and it is only their own property. This will considerably reduce the problems faced by the operation and maintenance staff of PWD and will result in efficient management of water resource system.

For successful implementation of the PIM, the following aspects need special attention. (a) Legal aspects - Appropriate Legislative backing should be provided as early as possible. (b) Institutional aspects - High Level Committee should be set up to formulate policies for the implementation of PIM and review policy issues from time to time. A Standing Committee for operational and monitoring purposes should be set up. The setting up of training and research institutions like WALMIS where they do not exist has been suggested. Employment of Community Organisers for motivating the farmers and later working with them in setting up WUAs has also been suggested.

Agricultural Growth and Sector Performance

The agricultural growth rate during the past five years has decelerated to about 1.5 per cent from about 3 to 3.5 per cent during the preceding 20 years, thus dropping below the population growth rate, for the first time during the past 40 years. Capital formation in agriculture at 1.3 per cent of the GNP is also one of the lowest in recent decades, adversely affecting irrigation and rural infrastructure development, whereas agriculture accounts for about 20 per cent of the national GDP and 60 per cent of the employment.

Technology fatigue in agricultural development is being felt widely. The total factor productivity growth rate particularly in the main Green Revolution belts of rice-wheat or rice-rice systems has decelerated. The problem of technological fatigue is further compounded with huge technology transfer gaps at various levels. Average national yields of most agricultural commodities in India are about 40 to 50 per cent of the corresponding World averages. The gaps between potential and realizable and between realizable and average realized yields in the country are generally around 50 to 100 per cent, respectively. The existing exploitable yield gaps should be seen as an opportunity for future growth that is consistent with agro-ecological, environmental, socio-economic, political and technological settings in the major production regimes.

Scope to Increasing Water Use Efficiency

Our Prime Minister has rightly emphasized the need to double annual foodgrain production from the present about 210 million tonnes to 420 million tonnes within the next 10 years. Since land is a shrinking resource for agriculture, the pathway for achieving these goals has to be higher productivity per units of arable land and water. Factor productivity will have to be doubled, if the cost of production is to be reasonable and the prices of our farm products are to be globally competitive. On an average, rice and wheat yields will need to be enhanced by about 40 per cent and pulses, oilseeds, maize, millets, sorghum and horticultural commodities yields by about 50 to 100 per cent.

Technological improvements in irrigation systems have also increased production opportunities. Traditional irrigation technologies (furrow, border and flood irrigation) which involve water delivery to plants through gravitation usually resulted in substantial water losses and limited uniformity in water distribution.

Modern irrigation technologies, particularly sprinkler and drip irrigation, increase water use efficiency. They have opened up opportunities to cultivate soils with low water-holding capacity (sandy and rock soils) and to farm low quality lands and steep slopes. This technology has also enabled regions facing limited water supplies to shift from low-value crops with high water requirements (*e.g.* cereal) to high value crops with lower water requirements such as fruits, vegetables and oil seeds. Water use efficiency is presently estimated to be only 38 to 40 per cent for canal irrigation and about 60 per cent for ground water irrigation schemes. On the basis of 1991 census, our country's per capita water availability per year was estimated at 2214 cubic metres against the global average of 9231 cubic metres. Irrigation, being the major water user, its share in the total demand is bound to decrease from the present 83 per cent to 74 per cent due to more pressing and competing demands from other sectors by 2025 A.D. and as such, the question of improving the present level of water use efficiency in general and for irrigation in particular assumes a great significance in perspective water resource planning.

It is estimated that with 10 per cent increase in the present level of water use efficiency in irrigation projects, an additional 14 m.ha area can be brought under irrigation from the existing irrigation capacities which would involve a very moderate investment as compared to the investment that would be required for creating equivalent potential through new schemes. There is considerable scope for preventing and alleviating drainage problems by more integrated planning and water management.

This can include integrated use of canal and groundwater for water table control, consideration of upstream and downstream relationships, adapting land use to the natural drainage conditions, exploration of opportunities of biological drainage and serial re-use of low quality drainage effluents. Both technology fatigue and technology gap should have no place in the Indian R&D system.

In fact, at this juncture, technology should flow faster through the pipeline and more options should be available to users, and agricultural research, education and extension systems should be revitalised. With the above backdrop, the researchers and technology developers thus must ask themselves the following questions in deciding their research and technology development priorities:

- Will the technology lead to higher productivity across all farms, water regimes (rainfed drylands), soil types and regions, not just large farmers and well-endowed ones?
- How will the technology affect the seasonal and annual stability of production, especially the highly risk prone rainfed areas suffering from high instability?
- How will the technology affect the energy balance, ecosystem and the sustainability of farming?
- Who will be the winners and losers from the technology - and how will it affect the majority small and marginal farmers, the poor and deprived ones?

Hence, performance of new varieties and technologies in terms of net income per hectare, and not just in terms of yield per hectare is important. Therefore, the aim of technological transformation of farming systems should be to enhance income per hectare on an environmentally and socially sustainable basis.

WATER LOGGING TROUBLE IN INDIA

National commission on Agriculture estimated that there is about 60 lakh in India suffering from improper drainage. Out of this, nearly 34 lakh hectares get inundated and the rest 26 lakh hectares has too high water table. Most of the low lying areas of the country have become unproductive due to water logging and salt accumulation. These lands generally have high production potential and require to be reclaimed by installing properly designed drainage system and adopting proper soil management practices.

The problem of improper drainage is especially serious in northern states like West Bengal, Punjab, Haryana, Rajasthan and Uttar Pradesh. The problem in these states has originated due to floods in rivers and faulty irrigation methods. The states of Andhra Pradesh, Orissa, Kerala, Tamil Nadu, Karnataka, Jammu and Kashmir, Himachal Pradesh and Assam face problems of inundation in rainy season.

Generally, southern plateau does not face serious drainage problem because of its undulating topography and the resultant tendency to surface drain the excess water. It is in the irrigated and low lying lands that this problem is serious. The states of Punjab and Haryana have the largest area affected by water logging which was estimated to be approximately 10.9 lakh hectares in 1958. About half of the area was affected by excessive salt problems. A large number of drainage schemes were carried out which consist essentially of constructing a network of drainage canals to dispose off the excess run-off. The extent of water logging in Uttar Pradesh is estimated to be about 8.1 lakh hectares.

Causes of Water Classification

The causes of water classification in different regions of the country can be attributed to the following factors.

- In the Indo-Gangetic plain, the main reason for water logging is flood in the rivers. Rainfall in this region is very uncertain in quantity and distribution. More than 90 per cent of the precipitation occurs in a short time span (within few months, *i.e.*, June to August). This results in occurrence of floods. Rain water is stored in low lying areas and deteriorates soil in the long run.
- In the last three-four decades, there has been a rapid increase in number of roads, highways, railway lines and city establishments. Inadequate attention has been given to construct drainage structures like culverts and drainage water ways. This resulted in obstructions to rationally flowing water and stagnation of water in depressed spots.
- With the increase of population pressure on land is also increasing. The land to receive drainage water is shrinking very rapidly.
- An extension network of canals was created in the ftrst half of the

present century, to meet the irrigation requirement of the arable lands. As the water become plentifully available, farmers started growing crops of high water requirements like paddy and sugarcane. This resulted in heightening of ground water table. Farmers used water unscrupulously contributing to the drainage problem. In those canal zones where soil is permeable, water table increases rapidly. Soluble salts also start moving towards soil surface along with seepage water. Soil become saline in the course of a few years. Canal induced water logging is a problem of mainly North India.

- The soil having hard kankar pan in subsoil and heavy soils are prone to flooding. In such soils, downward movement of water (infiltration) is very slow and water tend to accumulate on the surface.

MEASURES ADOPTED AGAINST WATER CLASSIFICATION

Punjab took lead by constituting a Water logging Enquiry Committee in 1925 to study and report the extent and causes of water logging which had assumed serious proportions in the irrigated areas and to indicate preventive measures. Various auxiliary schemes were also initiated and the total area benefited by these schemes is estimated to be about 1.6 lakh hectares. In Haryana, measures began with execution of the Jagadhari tubewell scheme which was aimed at pumping out water from water-logged areas and supplying this water to western Jamuna Canal.

Extensive drainage works have been undertaken in Gujarat after investigations on water logging and salt problem under Mahi Canal Project. Other states have also been carrying out drainage works and such schemes are providing flood embankments to aim of the schemes is to facilitat the removal of excess monsoon run-off.

EFFECT OF SHOCKING DRAINAGE ON SOIL AND PLANT

In the soil, water beyond field capacity is excess or redundant or useless for the plans. If excess water accumulates in the rootzone of the soil for a long time plants suffer badly. At this level of soil moisture, all the capillary pore spaces are packed with water and air is completely driven out of the porespaces. Soil structure deteriorates and lead to further reduction in oxygen movement in the rootzone of the soil.

This results in the following occurrences:

- A proper air-water balance of the pore space is disturbed. Oxygen content of the rootzone is reduced with corresponding increase in carbon dioxide content. In this situation, permeability of root hair to water is reduced. Plant suffers water scarcity though the soil is flooded with water.

- In oxygen deficient condition, certain plant nutrients become toxic to plants. For example, sulphur is converted into hydrogen sulphide gas which is toxic to plant roots.
- Soil organisms play an important role in building soil fertility. Due to reduced oxygen supply their activity is badly affected.
- The incidence of root rot disease is increased. For example, cereals, grape, santra and sugarcane are affected by root-rot, custard apple by fusarium rot and strawberry by black rot.
- Growth of root is hampered, when flooding is followed by a dry season, plants die very soon.
- Temperature of the soil goes down thereby adversely affecting physical, chemical and biological changes in the soil.
- The nutrients applied through fertilizers reach down into the deep soil and plan~ fail to avail these nutrients.
- When water evaporates from the surface, it leaves back salts. Water coming up from deep soil through seepage also brings soluble salts to the surface along with itself. Salts go on accumulating on the surface soil and soil becomes saline in the course of time.
- Soil structure is disturbed. Most of the crops, except paddy cannot be grown successfully on such soils.
- Water logging is deleterious to standing crop. Besides, field preparation is also delayed resulting in delayed sowing and reduced yield. Thus, both the crops are affected.
- The cattle grasing on inadequately drained areas may suffer from diseases like foot rot, chills, udder trouble and parasites.
- Unwanted vegetation like marsh grass, hedge and swamp trees come up.

DRAINAGE EVENTS

A plethora of problems arising due to frequent occurrence of floods, heightening water table, widening salinisation, etc., force us to think about the effective measures of preventing losses and rehabilitating already degraded lands due to these problems. Broadly, drainage measures may be divided into two main categories namely preventive measures and remedial measures.

Preventive Events

The main objective of these measures is to check the expansion of the problem. The main preventive measures are as follows:

Precaution by Farmers

If scientific use of irrigation water is a boon, it is necessary to use only as much water for irrigation as is just enough to meet the need of crop. This very much depends upon the type of soil and crop. It is worthwhile to mention here that irrigation water contains certain amount of salts. Usually this amount varies from 200 to 500 ppm. If the salt content is 200 ppm one irrigation of 8 ha cm will supply about 250 kg salt to a hectare land. The advise of the expert on amount of irrigation should be taken if the sources of irrigation water are known to be salty.

Technique of Irrigation

For different kind of crops, different methods of irrigation are recommended. These methods are developed by the experts keeping drainage aspect into account. Therefore, it is very useful from the drainage view point to follow recommended methods of irrigation for various crops.

Breaking Hard Pan

Now and then a bard clay pan develops in the sub-soil which hampers downward movement of water through soil and water tend to accumulate on the soil surface. Breaking hardpan by deep ploughing or chiseling increases water infiltration and reduces waterlogging. Excess of soluble salts- also reach down along with infiltrating water.

Avoidance of Seepage

A large amount of water seeps down from canals rivers and drains and get mixed with the ground water. Ground water table of surrounding areas is increased and problem of drainage sets in. Care should be taken while constructing these structures, to avoid seepage loss of water from drains and adequate increase in ground water table.

Small and short drains should be plastered with clay, canal beds should be compacted and spread with *kankars* and pebbles. Canal water should not be allowed to flow continuously, intermittent cease to the flow should be given. Accessory drains of canals should be dug parallel to canal and they should not be left blind, rather they should be connected with some pond or reservoir.

Remedial Events

Remedial events include those measures which are employed for reclaiming problem areas. Some important remedial measures are discussed below.

Wet Farming Practice

Wet farming is a combination of farming practices specially designed for growing crops in wet regions. This is suitable for the places where waterlogging is not very serious problem and there is a scope to reclaim the land.

Selection of Crop and Variety

Certain crops have genetic strength to resist waterlogging. The roots of such plants develop certain devices to capture oxygen when oxygen supply is scanty. Waterlogging resistance power of a plant also depends upon the duration of waterlogging growth stage of plants and the temperature. Cereals are usually more resistant than legumes to waterlogging. Among cereals maize is the least resistant. bajra is medium and jowar is the most, while among pulses lobia is the most resistant.

With each crop some varieties are more resistant than other. For example Ganga-2 variety of maize is more resistant to waterlogging than other varieties. Crops and varieties are sensitive to waterlogging at the time of flowering and grain formation.

Sowing on Ridge

In the areas where water-logging is shallow enough to allow construction of ridges raised above the water level seeds should be sown on ridges. In such condition roots of the seedling get enough oxygen to meet their requirement and incidence of diseases is also reduced significantly. In an experiment arhar, urd and mung yielded 30.7 per cent, 48.64 per cent and 58.5 per cent more respectively when they were grown on ridges in comparison to plain sowing.

By raising the ridges, furrows are automatically created which may be used for draining out excess water. Even though water remains ponding in the furrows, raised portion of ridges provide enough way for oxygen to reach the plant roots. Thus the yield can be increased significantly by sowing the seeds on ridges rather than in plain land.

Utilise of Fertilizers

The response of fertilizer use in wet fields is usually high. In such fields moisture is not a problem. Fertilizer nutrient dissolves in water and easily reaches inside the plant roots. A fast vegetative growth of plant takes place resulting in increased requirements of moisture and nutrients. Ultimately it results in increased yield of crop.

Alteration in Sowing Time

The sensitivity of a crop to waterlogging is highest at particular growth stage. Alteration in sowing time, in order to avoid simultaneous occurrence of sensitive growth stage and flooding, is found useful. This way, the harmful effects of flooding on crops can be avoided. However, this practice needs data of rainfall of 10 -20 years past. Based on these data information, prediction on current year rainfall, modification in the sowing time can be made accordingly. Crops are sensitive to flooding generally at nursery stage, pre-flowering stage, flowering stage and grain formation Stage.

RIVER DIVERSIONS AND REGIONAL CONFLICTS OVER WATER

Large dams are constructed for allowing major diversions of water from the natural drainage flow of the river. These diversions result in a major change in the distribution patterns of water in a basin, especially when they involve inter-basin transfers. They therefore generate new conflicts over the distribution of water between different regions. Regional conflicts become inter-state conflicts, and are rapidly enmeshed in inter-state and centre-state politics. The Telugu Ganga Canal, which takes off from the Srisailam Dam, is probably the most conflict-ridden river diversion project in contemporary India.

Krishna is the second largest river of peninsular India. Its catchment lies in the Western Ghats and it flows east through the states of Maharashtra, Karnataka and Andhra Pradesh. Krishna is an inter-state river, and conflicts have arisen between the co-riparian states over the allocation of its waters to their respective territories for purposes of development. The Krishna basin like other regions of India had indigenous irrigation works such as tanks, wells and anicuts. There were nearly 27,000 small tanks and diversions on the Krishna river system, mostly in Andhra Pradesh and Karnataka. To these were added new canals during the colonial period for commercial agriculture. These were:

1. The Krishna Delta Canals built in 1855.
2. The Nira Canals in Maharashtra constructed in I885 irrigating abou 150,000 acres.
3. The Kurnool Cuddapah Canal in Andhra Pradesh built in 1886 irrigating 100,000 acres.

The majority of the area irrigated by the Krishna Delta Canals and Kurnoo Cuddapah Canal (1.11 million acres) lies outside the basin of the Krishna. In 1951, the status of the diversion of Krishna waters was as follows: 411.4 TMCF of water was diverted annually for the irrigation of 2,302,377 acres. Of thi 290.1 TMCF was used by Andhra Pradesh, 430 TMCF by Maharashtra, and 78.3 TMCF by Karnataka.

After independence, the large-scale diversion of river waters increased In July 1951 the Planning Commission convened an inter-state conference t discuss the utilisation of Krishna waters. The dependable annual flow in the Krishna basin based on the recorded gaugings at Vijayawada was agreed a 1,715 TMCF so the balance of flow for new projects remained 970.5 TMC which was rounded off to 1,000 TMCF and allocations were made betwee the different states as follows:

For the balance flow in excess of 1,000 TMCF, if any, the allocation fo the above states was in the ratio 30:30:1:39. The state of Bombay was allowe to divert the waters to the west across the Western Ghats for the hydro-electr project at Koyna up to a limit of 67.5 TMCF. The agreement provided for review of the allocations after twenty years. In 1953, states were; reorganised on a linguistic basis, Madras was divided into Andhra and Madras. In 1956 the state of Andhra Pradesh was created by the merger of parts of Hyderabad and Andhra. As a result of territorial changes, the riparian states sharing the Krishna waters are Maharashtra, Karnataka and Andhra Pradesh. An interstate conference was convened in New Delhi under the auspices of the Union Minister of Irrigation and Power on September 1960, to recast the allocations of Krishna waters made in 1981. However, efforts to reach an agreement among the states proved unsuccessful and widely divergent views were expressed by the different states. ~ three man commission headed by N.D. Gulhati was set up. The commission undertook the first ever attempt 'at a basin-wide survey of the technical implementation relevant to water resources development.'

As observed by Tripathi, the Commission in examining the river flow of both these rivers was greatly hampered by the lack of regular reliable and continuous observations of water discharge at various points in river... The Commission stated that the flow records prior to 1936 were based on formulae different from those followed after 1936. Therefore the Commission stated that it [was] not possible to determine the flow for 86 per cent dependability or for 75 per cent dependability or for any other criterion of dependability. Because of the lack of adequate data of river flow, the Commission could not give positive answers to the terms of reference as regards the availability of water supplies on the river systems. State-wise allocation of Krishna waters was, therefore, not possible owing to the lack of scientifically observed data. The Irrigation Minister decided that adequate river data should be collected over a number of years and analysed continuously. However, tentative allocation was made for ongoing projects.

In spite of interim re-allocations, conflicts over Krishna waters continued with each state accusing the other of higher withdrawals from the river than its legitimate share. Maharashtra and Karnataka wanted a tribunal set up under Section 3 of the Interstate Water Disputes Act, 1956. The Bachawat Tribunal was appointed in 1969 to resolve the Krishna water state conflicts. In 1973 the Bachawat Committee gave its award. The availability of water was assessed at 2,060 TMCF and on the basis of 75 per cent dependability, Andhra Pradesh was allocated 800 TMCF.

The award fixed a formula for sharing both during surplus and lean years and was binding on the states until AD 2000. Mrs. Gandhi the then Prime Minister consulted the co-riparian states to provide drinking water to Madras which had been facing severe shortages. The three states readily agreed to part with S TMCF each. The Chief Minister of Andhra Pradesh later hailed the Krishna water supply scheme to Madras as the Telugu Ganga. The agreement was reached on 14 April 1976. On 17 October 1977, it was agreed that a 330 km long open canal would carry 15 TMCF to Madras.

While the decision to supply drinking water to Madras was agreed by all the states, conflicts arose when Andhra Pradesh decided to use the Telugu Ganga project for irrigation. The ₹ 850 crore project now envisages extension of irrigation to 5.75 lakh acres in three districts of Rayalseema-Kurnool, Cuddapah and Chittoor and one district in the Andhra region-Nellore, in addition to the supply of 15 TMCF of drinking water to Madras. Nearly 42.4 per cent of Kurnool district lies in the Krishna basin.

Cuddapah and Chittoor as well as the rest of Kurnool lie in the Pennar basin. Karnataka had questioned the diversion of water outside the basin to the KWDT arguing that only in-basin needs should be considered in determining a state's equitable share, a state should be permitted to divert its share of water outside the basin. Andhra Pradesh maintained that out of basin needs are a relevant factor and that diversions outside the basin for irrigation needs only should be permitted. Using precedence from the American Law, the Bachawat Tribunal held that the diversion of Krishna water outside the basin was legal. The river basin as an integral unit was thus substituted by the state as an administrative unit. The conflicting demands and distributive patterns emerging from the integrity of the basin versus the integrity of the state ifs a major reason for inter-state conflicts between riparian states not getting fully resolved.

Another fundamental reason for the intractable nature of river conflicts arises from the rights established through the priority of Project use in time and the rights based on the priority of need in the long term. Andhra Pradesh contends that the diversion of an additional 275 TMCF of Krishna waters to feed the districts of Kurnool and Cuddapah in Rayalseema for irrigation of 2.75 lakh acres, is within the scope of the Bachawat award as the Tribunal permitted Andhra Pradesh to take advantage of the surplus flows down the river at Vijayawada. As the Bachawat Tribunal stated, 'the state of An&a Pradesh will be at liberty to use in any year the remaining water that may be flowing in the Krishna river'.

Karnataka has objected to the Telugu Ganga irrigation scheme on the grounds that its own projects to harness Krishna waters are still incomplete and what appears to be excess, currently, will be used in the future. Karnataka has made it clear that surplus Krishna waters would not be available for the Telugu Ganga project. Maharashtra has also opposed the Telugu Ganga project on the ground that it violates the inter-state agreement reached in October 1977. The government of Maharashtra has observed that the state has vast chronic drought affected areas. Almost 75 per cent of the Krishna basin area in Maharashtra is drought prone and the state has plans to use the Krishna water allocated to it by the Krishna Tribunal. It has, therefore, to make sure that at the time of review of the award, its legitimate claim to the surplus available water in the Krishna river is not in any way jeopardised by pre-emptive efforts to commit this surplus water to projects like the Telugu Ganga.

Karnataka- and Maharashtra governments are resisting the project on the grounds that Andhra Pradesh has already used its allocation and the Telugu Ganga project would enable Andhra Pradesh to establish its right on larger volumes of water through prior utilisation.

Andhra Pradesh has already invested ₹ 200 crores and has 5,000 labourers working on the construction of the canal. Of the 406 km length of the canal, 190 km pass through the reserved forests of the Nellamali Range, for which central environmental clearance has not been obtained so far. At present the work is confined to reservoirs and canals in the non-forest areas. Water for the project is to be drawn from the Srisailam Dam through the head regulator at Pothireddypadu, which has a total carrying capacity of 11,000 cusecs.

The first 16 km of the canal is shared with the Srisailam right branch canal. The common canals run up to Bankacherla cross regulator where the Srisailam right branch canal and the Telugu Ganga Canal branch off to the right and left, respectively. The Telugu Ganga Canal is in fact the old Srisailam left branch canal extending into the Segileru Valley. Water to be drawn for the Telugu Ganga project is to be stored in four reservoirs at Yellgodu, Brahamasagar, Somashila and Kandaleru. At Mithakanda? near the Bankacherla regulator, a 100 feet high ridge divides the Krishna and the Pennar basins, where the water would be transferred outside the Krishna basin. The canal would pass through Kurnool and Cuddapah districts from where the water would flow into the Pennar river at Chenumukapalli. The flow down the river would be picked up at the Somashila Dam and passed on to the Kandaleru reservoir before it reaches Madras. The construction work continues even though the controversy over the Telugu Ganga project remains unresolved.

Andhra Pradesh derives its legitimacy from two arguments. First, it claims it is using only surplus waters for the project, and the right to surplus waters had been granted to it by the Tribunal. Second, it claims that if there is scarcity, then the arid drought prone regions of Rayalseema should not be asked to sacrifice irrigation waters. Instead, Maharashtra should be asked to stop diverting large volumes of water out of the Krishna basin into the Arabian Ocean for power generation for industrial centres. The river Krishna emerges in the Western Ghats and flows eastward down the gentle slopes. The western face of the Western Ghats falls steeply down altitudes of 1,000 to 2,000 feet, providing excellent sites for power generation. However, the water used for hydro-electricity has to be diverted out of the basin, and dropped into the sea. Currently, the power projects in Maharashtra which divert water westwards are the Tata and Koyna Hydel Projects. The former diverts 42.6 TMC and the latter diverts 67.5 TMC.

In this conflict between the demands for power generation and the demands for irrigation in drought prone areas, the Krishna Tribunal protected existing diversions while giving priority to irrigation for future use. As it stated: In the Krishna Basin, water is a scarce commodity. Westward diversion of water for power generation seriously restricts the use of water for downstream irrigation.... Power for Bombay and Maharashtra industry is generated at the cost of depriving the low rainfall areas on the eastern side of the water solely needed for irrigation.

The Tribunal, however, allowed the expansion of hydel projects on the condition that over a period of twenty years they would return to the existing capacity. When the next Krishna Tribunal meets in the year 2020 to review the sharing and utilisation of Krishna waters, the concepts of justice and rights as related to water will have undergone dramatic changes, as will the basin itself.

8

Basic of Drainage

INTRODUCTION

Drainage can be either natural or artificial. Many areas have some natural drainage; this means that excess water flows from the farmers' fields to swamps or to lakes and rivers. Natural drainage, however, is often inadequate and artificial or man-made drainage is required. There are two types of artificial drainage: surface drainage and subsurface drainage.

Surface Drainage

Surface drainage is the removal of excess water from the surface of the land. This is normally accomplished by shallow ditches, also called open drains. The shallow ditches discharge into larger and deeper collector drains. In order to facilitate the flow of excess water towards the drains, the field is given an artificial slope by means of land grading.

DRAINAGE IN THE 19TH CENTURY

This operation is always best performed in spring or summer, when the ground is dry. Main drains ought to be made in every part of the field where a cross-cut or open drain was formerly wanted; they ought to be cut four feet (1.2 m) deep, upon an average. This completely secures them from the possibility of being damaged by the treading of horses or cattle, and being so far below the small drains, clears the water finely out of them. In every situation, pipe-turfs for the main drains, if they can be had, are preferable.

If good stiff clay, a single row of pipe-turf; if sandy, a double row. When pipe-turf cannot be got conveniently, a good wedge drain may answer well, when the subsoil is a strong, stiff clay; but if the subsoil be only moderately so, a thorn drain, with couples below, will do still better; and if the subsoil is very sandy, except pipes can be had, it is in vain to attempt under-draining the field by any other method. It may be necessary to mention here that the size of the main drains ought to be regulated according to the length and declivity of the run, and the quantity of water to be carried off by them. It is always safe, however, to have the main drains large, and plenty of them; for economy here seldom turns out well. Having finished the main drains, proceed next to make a small drain in every furrow of the field if the ridges formerly have not been less than fifteen feet (5 *m*) wide. But if that should be the case, first level the ridges, and make the drains in the best direction, and at such a distance from each other as may be thought necessary.

If the water rises well in the bottom of the drains, they ought to be cut three feet (1 *m*) deep, and in this ease would dry the field sufficiently well, although they were from twenty-five to thirty feet (8 to 10 *m*) asunder; but if the water does not draw well to the bottom of the drains, two feet (0.6 *m*) will be a sufficient deepness for the pipe-drain, and two and a half feet (1 *m*) for the wedge drain.

In no case ought they to be shallower where the field has been previously levelled. In this instance, however, as the surface water is carried off chiefly by the water sinking immediately into the top of the drains, it will be necessary to have the drains much nearer each other--say from fifteen to twenty feet (5 to 6 *m*). If the ridges are more than fifteen feet (5 *m*) wide, however broad and irregular they may be, follow invariably the line of the old furrows, as the best direction for the drains; and, where they are high-gathered ridges, from twenty to twenty-four inches will be a sufficient depth for the pipe-drain, and from twenty-four to thirty inches for the wedge-drain. Particular care should be taken in connecting the small and main drains together, so that the water may have a gentle declivity, with free access into the main drains.

When the drains are finished, the ridges are cleaved down upon the drains by the plough; and where they had been very high formerly, a second clearing may be given; but it is better not to level the ridges too much, for by allowing them to retain a little of their former shape, the ground being lowest immediately where the drains are, the surface water collects upon the top of the drains; and, by shrinking into them, gets freely away. After the field is thus finished, run the new ridges across the small drains, making them about ten feet (3 *m*) broad, and continue afterwards to plough the field in the same manner as dry land. It is evident from the above method of draining that the expense will vary very much, according to the quantity of main drains necessary for the field, the distance of the small drains from each other, and the distance the turf is to be carried.

The advantage resulting from under-draining, is very great, for besides a considerable saving annually of water furrowing, cross cutting, etc., the land can often be ploughed and sown to advantage, both in the spring and in the fall of the year, when otherwise it would be found quite impracticable; every species of drilled crops, such as beans, potatoes, turnips, etc., can be cultivated successfully; and every species, both of green and white crops, is less apt to fail in wet and untoward seasons. Wherever a burst of water appears in any particular spot, the sure and certain way of getting quit of such an evil is to dig hollow drains to such a depth below the surface as is required by the fall or level that can be gained, and by the quantity of water expected to proceed from the burst or spring. Having ascertained the extent of water to be carried off, taken the necessary levels, and cleared a mouth or loading passage for the water, begin the drain at the extremity next to that leader, and go on with the work till the top of the spring is touched, which probably will accomplish the intended object.

But if it should not be completely accomplished, run off from the main drain with such a number of branches as may be required to intercept the water, and in this way disappointment will hardly be experienced. Drains, to be substantially useful, should seldom be less than three feet (1 *m*) in depth, twenty or twenty four inches thereof to be close packed with stones or wood, according to circumstances. The former are the best materials, but in many places are not to be got in sufficient quantities; recourse therefore, must often be made to the latter, though not so effectual or durable. It is of vast importance to fill up drains as fast as they are dug out; because, if left open for any length of time, the earth is not only apt to fall in but the sides get into a broken, irregular state, which cannot afterwards be completely rectified. A proper covering of straw or sod should be put upon the top of the materials, to keep the surface earth from mixing with them; and where wood is the material used for filling up, a double degree of attention is necessary, otherwise the proposed improvement may be effectually frustrated.

The pit method of draining is a very effectual one, if executed with judgment. When it is sufficiently ascertained where the bed of water is deposited, which can easily be done by boring with an auger, sink a pit into the place of a size which will allow a man freely to work within its bounds. Dig this pit of such a depth as to reach the bed of the water meant to be carried off; and when this depth is attained, which is easily discerned by the rising of the water, fill up the pit with great land-stones and carry off the water by a stout drain to some adjoining ditch or mouth, whence it may proceed to the nearest river.

CURRENT PRACTICES

Modern drainage systems incorporate geotextile filters that retain and prevent fine grains of soil from passing into and clogging the drain. Geotextiles are synthetic textile fabrics specially manufactured for civil and environmental engineering applications. Geotextiles are designed to retain fine soil particles while allowing water to pass through. In a typical drainage system they would be laid along a trench which would then be filled with coarse granular material: gravel, sea shells, stone or rock. The geotextile is then folded over the top of the stone and the trench is then covered by soil. Groundwater seeps through the geotextile and flows within the stone to an outfall. In high groundwater conditions a perforated plastic (PVC or PE) pipe is laid along the base of the drain to increase the volume of water transported in the drain. Alternatively, prefabricated plastic drainage systems made of HDPE called Smart-Ditch, often incorporating geotextile, coco fiber or rag filters can be considered. The use of these materials has become increasingly more common due to their ease of use which eliminates the need for transporting and laying stone drainage aggregate which is invariably more expensive than a synthetic drain and concrete liners. Over the past 30 years geotextile and PVC filters have become the most commonly used soil filter media. They are cheap to produce and easy to lay, with factory controlled properties that ensure long term filtration performance even in fine silty soil conditions.

21st Century Alternatives

Seattle's Public Utilities created a pilot programme called Street Edge Alternatives (SEA Streets) Project. The project focuses on designing a system "to provide drainage that more closely mimics the natural landscape prior to development than traditional piped systems".

The streets are characterized by ditches along the side of the roadway, with plantings designed throughout the area. An emphasis on non curbed sidewalks allows water to flow more freely into the areas of permeable surface on the side of the streets. Because of the plantings the run off water from the urban area does not all directly go into the ground but can also be absorbed into the surrounding environment. According to the monitoring by Seattle Public Utilities, they report at 99 per cent reduction of storm water leaving the drainage project.

DRAINAGE IN CONSTRUCTION

The civil engineer or site engineer is responsible for drainage in construction projects. They set out from the plans all the roads, Street gutters, drainage, culverts and sewers involved in construction operations. During the construction of the work on site he/she will set out all the necessary levels for each of the previously mentioned factors.

Site engineers work alongside architects and construction managers, supervisors, planners, quantity surveyors, the general workforce, as well as subcontractors. Typically, most jurisdictions have some body of drainage law to govern to what degree a landowner can alter the drainage from his parcel.

REASONS FOR ARTIFICIAL DRAINAGE

Wetland soils may need drainage to be used for agriculture. In the northern USA and Europe, glaciation created numerous small lakes which gradually filled with humus to make marshes. Some of these were drained using open ditches and trenches to make mucklands, which are primarily used for high value crops such as vegetables. The largest project of this type in the world has been in process for centuries in the Netherlands. The area between Amsterdam, Haarlem and Leiden was, in prehistoric times swampland and small lakes. Turf cutting (Peat mining), subsidence and shoreline erosion gradually caused the formation of one large lake, the Haarlemmermeer, or lake of Haarlem. The invention of wind powered pumping engines in the 15th century permitted drainage of some of the marginal land, but the final drainage of the lake had to await the design of large, steam powered pumps and agreements between regional authorities.

The elimination of the lake occurred between 1849 and 1852, creating thousands of km^2 of new land. Coastal plains and river deltas may have seasonally or permanently high water tables and must have drainage improvements if they are to be used for agriculture. An example is the flatwoods citrus-growing region of Florida. After periods of high rainfall, drainage pumps are employed to prevent damage to the citrus groves from overly wet soils. Rice production requires complete control of water, as fields need to be flooded or drained at different stages of the crop cycle. The Netherlands has also led the way in this type of drainage, not only to drain lowland along the shore, but actually pushing back the sea until the original nation has been greatly enlarged.

In moist climates, soils may be adequate for cropping with the exception that they become waterlogged for brief periods each year, from snow melt or from heavy rains. Soils that are predominantly clay will pass water very slowly downward, meanwhile plant roots suffocate because the excessive water around the roots eliminates air movement through the soil. Other soils may have an impervious layer of mineralized soil, called a hardpan or relatively impervious rock layers may underlie shallow soils.

Drainage is especially important in tree fruit production. Soils that are otherwise excellent may be waterlogged for a week of the year, which is sufficient to kill fruit trees and cost the productivity of the land until replacements can be established. In each of these cases appropriate drainage carries off temporary flushes of water to prevent damage to annual or perennial crops.

Drier areas are often farmed by irrigation, and one would not consider drainage necessary. However, irrigation water always contains minerals and salts, which can be concentrated to toxic levels by evapotranspiration. Irrigated land may need periodic flushes with excessive irrigation water and drainage to control soil salinity.

DRAINAGE TECHNIQUES

An effective drainage system is essential for a well managed upland path. If the drainage system does not function properly erosion scars become severe, and any path surface work can be destroyed after one winter of rainfall, if not less.

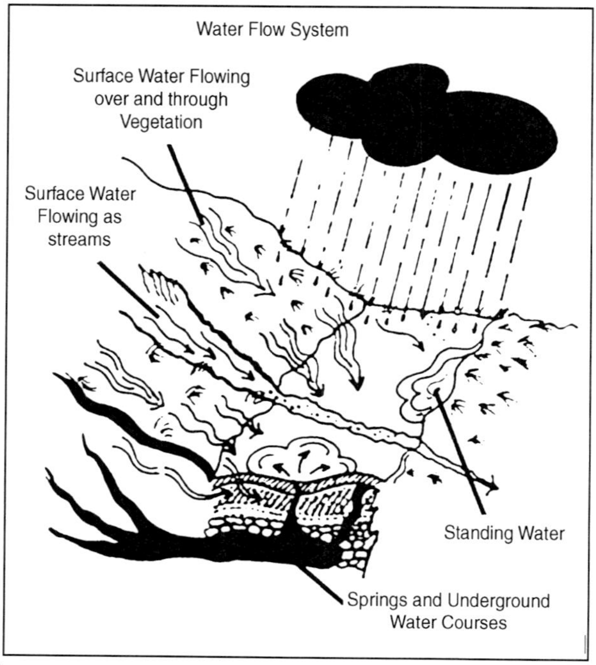

There are three main sources of water on the path:

- Rain falling directly onto and running down the path surface, or snow melting
- Surface water from surrounding land flowing directly onto the path
- Underground water running onto the path surface, in the form of springs or seepage.

An assessment of the climdte and altitude can give clues about these water patterns.

- Naturally high rainfall, *e.g.*, western mountain areas
- Altitude, and latitude, indicating the likelihood of snow cover
- Seasonal snow fall and problems associated with potentially sudden snow-melt

ASSESSING THE SITE

The site must be assessed before any decisions can be made about which drainage features are appropriate, how many are needed and the positioning to achieve adequate protection of the path. At the same time the impact of changing the natural drainage system must be considered, particularly in environmentally sensitive areas where natural vegetation of ecological value is dependant on a particular source or level of water.

Assessment, or specification survey should preferably take place on a wet day, or just after a particularly wet period of weather. If this is not possible clues will be found on the path and the surrounding landscape. The full length of the path should be walked, noting where water is coming from, both on the way up and on the way back down. These two perspectives should help to ensure that all aspects are considered. It is often the case that a drainage problem further up the path can be causing the problems below. For instance a stream at the top end of the path may have burst its banks or erosion debris diverted its course, directing flow straight down the path line.

The first clues come from the path itself. It should be possible to identify:

- Damage caused by water flowing down or across the path - evident as gullies or eroded channels.
- Damage caused by water lying on the path - evident as puddles, boggy areas often saturated with water, signs of walkers skirting round and causing braiding.
- Where water has come from - evident by signs of springs, surface water, or water flowing onto the path from the slopes above.
- Where water is going - evident by signs of silt at the path edge, or vegetation may show signs of being flattened by water flowing off the path.

The immediate landscape can also provide information about how the site reacts after periods of rain, and where and what drainage features are required:

- Geology of the area can indicate whether water flows close to the surface of the ground - evident by large areas of bedrock; or if it soaks away - evident as vegetated areas of deep soil or peat.
- Topography of the area can indicate where the water flows - quickly downhill if steep gradients are present and streams have formed; soaking away in flat areas of lush vegetation growth.
- Vegetation type can indicate areas of permanently wet ground by high presence of mosses, cotton grass or rushes. It can also show where water has flowed over it, flattening long grasses or deposited silt, inhibiting growth.

The Drainage Organisation

Having assessed the site, an appropriate and effective drainage system can be designed to combat the problems identified and protect the path.

There are two basic methods of drainage which are generally used in combination. Ditching is integral to both methods:

1. *Off path drainage:* to protect the path from water flowing onto it from the surrounding land.
   - Ditching intercepts the water before it reaches the path and drains it away
   - Culverts or cross-drains channel the intercepted water across the path
2. *On path drainage:* to divert water off the path surface.
   - Cross-drains collect water at low points and channel it away into ditches

- Water-bars deflect running water off the path
- Letts drain away puddles that have formed on the path

Design and use of different techniques should also take account of the existing or expected path users. For instance a boxed culvert is preferable to a cross-drain if a path is used by ATVs for deer management; a cross drain with a narrower channel is preferable to a water bar if there is use by cyclists.

PURPOSE OF DRAINAGE DESIGN

The purpose of drainage is to lower the water table far enough below the ground surface that it will not interfere with plant root growth. The degree of drainage required depends upon the maximum allowable height of the water table, the minimum rate at which the water table must be lowered, or the maximum allowable duration and frequency of ponding. The designer of the subsurface drainage system should select the degree of drainage that will fit the various crop requirements of the site.

Drain Size

One of the first steps in determining drain size is to select the drainage coefficient, which is the rate at which water is to be removed from an area. It is a value selected to provide adequate drainage for future crops and is expressed in inches per 24 hours. Watercourses provide adequate surface drainage, the drainage area for which you are choosing a drainage coefficient need only include the area that will be drained by subsurface drains. If the slope of the field is less than 0.2 per cent, choose the higher of the drainage coefficient ranges listed in the table.

Where surface drainage is not adequate and surface-water or blind inlets must be used to drain depressions, the drainage coefficient must be relatively high so that the drains can remove runoff from the entire watershed of the depressional area. An exception can be made where the depressions are small, as long as surveys are available and the volume of the potholes can be determined accurately. In that case, the drains should be able both to remove water at the appropriate drainage coefficient from the land area that needs drainage and to remove the water in the potholes within 24 to 48 hours. The size of the drain depends not only upon the drainage coefficient, but also upon the size of the area to be drained, the grade of the drain, and the internal roughness of the pipe.

The main should be large enough to drain all areas in the watershed that need drainage at the appropriate drainage coefficient. It should also have a free outlet and be deep enough to provide an outlet for all laterals to be installed. The interactive Sizing Drainage Pipes programme. First, find the appropriate drainage coefficient for the crop and drainage conditions. Next, locate the point at which a horizontal line through your acreage would intersect a vertical line through your drain grade or slope (horizontal axis). This point indicates the size of the drain and the velocity of water moving through it when it is flowing at capacity. The minimum cleaning velocity is 0.5 feet per second for drains not subject to the entry of fine sand or silt and 1.4 feet per second where fine sand or silt may enter. The rate of discharge information can be used in determining drain size at design slopes.

The smallest drain generally recommended for laterals is 4 inches. A drain 3 inches in diameter may be installed in locations where the grade is 0.2 per cent or more, or where the design velocity exceeds 1.4 feet per second. The drain should be a minimum of 5 inches in diameter for a system with short laterals in sandy soils.

Normally, 6 inches is the minimum diameter for drains located in organic soils and for main lines. For a subsurface system that contains tile lines exceeding 10 inches in diameter, it is preferable to use 2-foot or greater lengths to maintain alignment.

Drain Length

The length of lateral drains made of corrugated plastic drain tubing and concrete and clay drain, assuming that the drains are spaced 100 feet apart and that the drainage coefficient is 3/8 inch. To determine drain length for other drainage coefficients and lateral spacings.

Drain Grade and Velocity

When possible, subsurface drains should be placed at uniform depths. The range of grades on which they can be placed depends to some degree upon the topography of the land. The grade should be great enough to prevent silting but flat enough to prevent flow from exceeding the allowable velocity and subjecting the drain to excessive pressure.

Too much flow would cause erosion around the drain. The grade should be as great as possible on flatlands. But you should not sacrifice adequate drain depth to increase the grade.

Wherever the grades of drains are flatter than the minimum, take these precautions to reduce the amount of sediment.

- Make sure that the system has a free outlet so that backwater conditions will not further reduce velocity.
- Provide sediment traps and clean-out systems.
- Provide breathers and relief wells to vent the drain and to assure maximum flow.
- Protect the entire system from sedimentation by using filters and envelopes to prevent movement of the drain blinding materials.

For long laterals and main drains, the maximum velocity should be limited to those listed below, assuming that no protective measures are provided.

Soil Texture	Velocity (feet/sec)
Sand and sandy loam	3.5
Silt and silt loam	5.0
Silty clay loam	6.0
Clay and clay loam	7.0
Coarse sand or gravel	9.0

If protective measures do prove to be necessary, use one or more of the following.

For clay or concrete tile:

- Use tile uniform in size and shape with smooth ends.
- Lay the tile to secure a tight fit. The inside section of one tile should match that of the adjoining tile.
- Wrap open joints with tar- impregnated paper, burlap, or special filter material such as plastic sheets, fiberglass fabric, or properly graded sand and gravel.
- Select the least erodible soil for blinding.
- Tamp soil carefully under and alongside the tile before back- filling.
- Cement joints or use a drain with water-tight joints.

For corrugated plastic tubing or continuous pipe: Completely encase perforated drains with a filter material made of plastic, fiberglass, or a like material, or use a properly graded sand and gravel filter. Use non perforated corrugated plastic tubing or continuous pipe with taped or leak-proof connections.

Drain spacing and depth: The spacing and depth of drains influences the groundwater level between drains after a rain. Good drainage lowers the water table to at least 12 inches below the ground surface in the first 24 hours after a rain and to approximately 21 inches 48 hours after a rain. Incorrect spacing and depth could result in water remaining in the fields after 24 to 48 hours, significantly affecting crops.

The spacing and depth required to keep the water table at the desired level are influenced by the permeability of the soil, depth to the barrier, the amount and frequency of rainfall, see page, capillary movement, and topography. Spacing and depth also influence each other. In general, you should increase the lateral spacing the deeper you place the drain. Spacing and depth recommendations are given in the Drainage Guidelines section for specific soils in Illinois. For soils not listed in the guidelines, keep in mind the following general principles about drain spacing and depth.

Drains in rapidly permeable soils should be spaced 200 to 300 feet apart, while those in moderately rapidly permeable soils should be spaced 100 to 200 feet apart. Where soil permeability is moderate, spacing should be 80 to 100 feet apart. In slowly permeable or moderately slowly permeable soils, drains should be spaced 30 to 70 feet apart or 60 to 80 feet apart, respectively.

With respect to general principles for depth, drains in moderate to moderately permeable mineral soils in humid areas should be installed at a depth of 3 to 5 feet. At this depth the drains will lower the water table to not less than 2 to 4 feet. Because the upward capillary action is limited in very sandy soils, the drains should be no deeper than 4 feet. In slowly permeable clay soils, the rate of lateral water movement does not increase with depth. Therefore, the drain is usually placed approximately 1 foot below the desired water table.

Table. General Parallel Drain Lateral Spacing and Depths for Different Soils

Soil Type	Subsoil Permeability	Drain Spacing (ft) for: Fair Drainage	Good Drainage	Excellent Drainage	Drain Depth (ft)
		1/4 in	3/8 in	1/2 in	
Clay loam	Very low	70	50	35	3.0 - 3.5
Silty clay loam	Low	95	65	45	3.3 - 3.5
Silt loam	Moderately low	130	90	60	3.5 - 4.0
Loam	Moderate	200	140	95	3.8 - 4.3
Sandy Loam	Moderately High	300	210	150	4.0 - 4.5

The depth of drains also depends upon conditions other than those mentioned above - the depth of frost penetration, for example. If possible, place drains below the frost line to obtain optimum year-round drainage and to prevent damage to the line.

To protect a well-bedded sub-surface drain in a mineral soil from breakage or excess deflection of flexible tubing by heavy equipment, make sure that the drain has a minimum coverage of 2 feet. Where 3-inch drains are used both for drainage and sub-irrigation of shallow-rooted crops, the minimum depth may be 1.5 feet if heavy machinery is not used in the cropped area. The minimum depth of cover in organic soils should be 2.5 feet for normal field levels after initial subsidence.

If controlled drainage is not provided to hold subsidence to a minimum, the depth of cover should be increased to 3 feet. The outlet should be deep enough for the lateral drain to have adequate grade and cover. When it is impossible to provide minimum cover for protection, use metal or some other continuous high-strength pipe.

Plastic drain tubing should be installed in such a way that it does not deflect more than 20 per cent of its inside diameter. The maximum trench depths for tubing that is buried in loose. Because the maximum depths listed in the table are based on a limited amount of research, they should be used with caution. Keep in mind, too, that these values are based on certain assumptions about corrugation design and pipe stiffness (which may not be the same for all commercial tubing) and about soil conditions. To prevent overloading in deep and wide ditches, you may want to construct a subditch, either with a trenching machine or by hand, in the bottom of a wide ditch that has been excavated by a bulldozer, dragline, power shovel, or backhoe. It is now the width of the subditch measured at the top of the drain that influences the allowable load; the width of the excavation above that point is relatively not important. It is a good idea to backfill deep trenches in several stages to allow time for settlement between fillings.

Interceptor drains: An interceptor drain can be used in areas that are wet because of seepage from adjoining highlands. The drain is also used to intercept seepage or water that flows in a pervious layer on top of an impervious subsoil stratum.

Proper location of an interceptor drain is very important. An interceptor drain is usually buried at about the upstream boundary of the wet area. The drain is installed at approximate right angles to the flow of groundwater and intercepts a seep plane in the soil profile. Adequate field investigations must be made to determine the amount of seepage and to identify seep planes. You can locate seep planes by making backhoe test pits or taking soil borings.

Interceptor Drain

An interceptor drain must intercept the seep plane continuously, have adequate soil coverage, and be on a continuous grade toward the outlet. The drain is usually located one-half to one times its diameter deep in the impervious layer or seepage plane. One or two properly spaced interceptor drains will usually dry up a wet area.

The flow will often be continuous throughout much of the year. In a steeply graded depression or draw, the layout may consist of a main or submain located in the draw or to one side of it and interceptor lines located across the slope on grades slightly off contour. To determine the size of an interceptor drain for a particular set of conditions, refer to the list below, which contains inflow rates for various soil textures. Measure or estimate the discharge of flowing springs and the direct entry of any surface flow through a surface inlet or filter.

Add that figure to the rate of inflow. If the interceptor lines are being installed on sloping land, increase the inflow rate by 10 per cent for slopes of 2 to 5 per cent, by 20 per cent for slopes of 5 to 12 per cent, and by 30 per cent for slopes of more than 12 per cent. The Sizing Drainage Pipes programme For interception areas where there is considerable seepage, the minimum drain size should be 6 inches.

Soil Texture	Inflow rate per 1,000 feet of line (cu ft/sec)
Sandy loam	0.07 to 0.25
Silt loam	0.04 to 0.10
Clay and clay loam	0.02 to 0.10
Coarse sand or gravel	0.15 to 1.00

Changing the Direction of Drains

You can change the horizontal direction of drain lines by several means. Curve the trench gradually on a radius of curvature that the trenching machine can dig while still maintaining grade. The joint spacings for tile should be no more than 1/16 to 1/8 inch. Use manufactured bends or fittings, or use junction boxes where drain lines make an abrupt change in direction or where two or more large drains join.

DRAINAGE MATERIALS

Drainage materials should be protected from damage during handling and storage. The storage area should be dry, well drained, and free of rodents, vegetation, and fire hazards. It should have a protected floor (peagravel or cement) and adequate security. Take more precautions to protect plastic tubing. Where rodents could be a problem, we recommend that you use end caps. Tubing with filter wrap should either be stored inside or placed in protective bags. Since tubing can be harmed by excessive exposure to ultraviolet rays, protect it from sunlight when it is to be stored outside for a long period. To protect coils or reels of tubing from damage and deformation, lay them flat when they are stored for extended periods. Coils of tubing should be stacked no more than four high; reels should not be stacked.

Tree Removal

Before installation begins, remember to remove willow, elm, soft maples, cottonwood, and other water-loving trees that grow within approximately 100 feet of the planned drainage lines. All other species of trees, with the possible exception of fruit trees, should be removed for a distance of 50 feet. If it is not possible to remove trees or to reroute the line, use a non perforated line with sealed joints throughout the root zone of the tree or trees.

Grade Control System

All drains should be installed at a predetermined grade that will give them the capacity required for the area to be drained. This grade should be maintained constantly during installation. The easiest way to control the grade is to use equipment that is designed especially for drainage installation. If you use a backhoe or other equipment, you must take extra precautions to ensure that the trench is shaped properly and the grade is maintained. Grade is normally maintained by the use of electronic, optical grade control devices (lasers). Small and gradual variations from grade can be tolerated, providing the line still has adequate capacity after the variations. No reverse grade should be allowed. If the trench is excavated below the designed grade, it should be filled to grade with gravel or well-pulverized soil and tamped enough to provide a firm foundation. The bottom of the trench should then be planed and shaped to grade.

Safety Standards

Observe safety standards for persons and machines. Persons working in trenches should be protected from cave-ins, and they should not work alone. Moving parts of machinery should be protected by proper guards. Persons observing the work should not be permitted to come close to the excavating operation.

TRENCH METHOD OF INSTALLATION

Constructing the Trench

Construction of the trench should begin at the outlet and proceed upgrade. Align trenches in such a way that the drain can be laid in straight lines or in smooth curves. The width of the trench at the top of the drain should be the minimum required to permit installation and enable the bed to support the load on the drain. But there should be at least 3 inches of clearance on either side of the drain.

Tile should be bedded in an earth foundation that is shaped to fit the lower part of the tile. The foundation can be shaped in this way with most trenching machines. If you dig the trench with a backhoe, you will have to hand grade and shape the trench bottom to fit the tile. For corrugated plastic tubing, a specially shaped groove must be made in the trench bottom if the design does not call for a gravel envelope. The groove provides side and bottom support to the lower part of the tubing and provides a means of controlling alignment during installation. The groove may be in the shape of a semicircle, trapezoid, or a 90-degree *V*. A 90-degree *V* groove of sufficient depth is recommended for 3- to 6-inch tubing. However, if the tubing is installed on a steep grade, shape the bottom of the trench to closely fit the tubing.

The groove for tubing can be formed or cut in a number of ways. With all methods, some type of a forming tool is attached to the shoe of the trenching machine. One method is to install a forming tool on the bottom of the shoe and use pressure to form the bedding groove. Another method is either to install a device on the front of the finishing shoe that will plow out the groove during the trenching operation or to attach special shaping cutters to the trenching wheel. The latter methods minimize soil compaction and do not reduce permeability as much as the first method.

The dimensions of a 90-degree V groove. The depth at which 3- to 6-inch tubing is set in the groove will vary according to the size of the tubing. This depth can be evaluated with the V-Groove programme. Keep this depth in mind when setting the gradeline. For tubing 8 inches or more in diameter, we recommend a curved bottom that more nearly fits the tubing rather than the 90-degree *V* groove. If the drain is to be laid in a rock-cut, the trench should be overexcavated to a depth of 6 inches below grade level; this space should be filled with graded sand and gravel or well-pulverized soil and tamped enough to provide a firm foundation. Then, the bottom of the trench should be shaped and levelled to grade. The trench should be filled with designed bedding or envelope material to the top of the rock-cut. Where the trench bottom is unstable, as in fine sandy soils or in soils containing quicksand, be extremely careful to keep sediment from entering the drain and to provide a firm foundation for the drain.

When draining these types of soils, consider the following suggestions:

- Install the drain only when the soil profile is in the driest possible condition.
- Place stabilizing envelope materials under the drain.
- Cover the remainder of the drain with an envelope material.
- Use non perforated tubing, self-sealing sewer pipe, or continuous rigid pipe where there are small pockets of noncohesive soils less than 100 feet in length.
- If phe drain is tile, be sure that the joints are snug. If tubing is used, take precautions to prevent it from floating.
- If you find unstable soil at the trench bottom, you can remove and replace it with suitably graded foundation and bedding of processed stone or processed gravel, which will act as an impervious mat into which the unstable soil will not penetrate. The depth of the processed material depends on how unstable the soil is in the trench bottom. Install the foundation and bedding material in layers of no more than 6 inches, and compact. If the foundation contains large particles that create a hazard to the drain, provide a cushion of acceptable bedding material between the foundation and the drain.

Where stabilizer materials do not furnish adequate support, the drain should be placed in a 90-degree, rigid *V*, prefabricated foundation cradle in which the top of the *V* equals the outside diameter of the drain. Each section of the cradle must provide rigidity and continuous support throughout the entire length of the cradle. Occasionally, it is necessary to place the cradle on piling; drive pairs of posts along the edge of the cradle into solid material to provide the required support.

If the soil in the trench wall is unstable, the trench sidewalls may cave in and cause tubing failure. This problem may arise where excavation is below groundwater level or in saturated sand. Unstable trench walls may also cause misalignment of tile lines. Where there are unstable trench sidewalls, you should protect the tubing or tile by some means until the drain has been properly laid and blinded. In some cases, the trencher shield behind the shoe can be made longer to protect a greater length of the trench during construction. To install drains in unstable soils, use a fast-moving trencher or trench less drain plow that can maintain continuous forward motion while disturbing the soil as little as possible.

Installing the Drains

Listed below are some guidelines to follow when installing drains:

- Remove all soil or debris inside drains before Installation.
- Make sure the drain is free from clinging wet or frozen Material that could hinder laying the drain on grade.
- Begin laying tile or tubing at the outlet and progress Upgrade. If possible, place the drain inside the shoe casing of the trencher
- During the trenching operation.
- Automatic drain-laying devices are acceptable, provided That they can lay the drain according to the requirements stated in this Publication.
- Lay tubing in the groove and tile on a firm bed that is free of loose soil on the planned grade Hold plastic tubing in position on grade immediately after Installation by careful placement of blinding material.
- Where lengths of plastic tubing are to be joined, cut the Ends square and remove all ragged or burred edges. Use a plastic coupling to
- Secure the ends of the tubing in proper alignment and to prevent the joint from Separating during installation.
- Before work is suspended for the day, blind and backfill all drains laid in trenches.
- Close any open ends tightly with an end plug.

Use continuous pipe within 100 feet of trees. Any stretch that occurs during installation of tubing will decrease its strength somewhat and may pull perforations open wider than is desirable. The amount of stretch that occurs during installation depends on the temperature of the tubing at the time it is installed, the amount and duration of drag that occurs when the tubing is fed through the installation equipment, and the stretch resistance of the tubing. Tubing should not be stretched so much that its stiffness is reduced to less than the minimum allowable pipe stiffness. Stretch, which is expressed as a percentage increase of length, should not exceed 5 per cent. The use of a power feeder is recommended for all sizes of tubing.

The internal wall temperature of plastic tubing can reach 1500 *F*. or more when it is strung out in a field on a hot, bright day. The ability of corrugated polyethylene tubing to resist deflection is reduced by about 40 per cent when its temperature rises from 700 to 1000 *F*. and by about 50 per cent when it increases from 700 to 1200 *F*. Therefore, it is essential that the contractor take precautions in hot weather to keep sharp heavy objects from striking the tubing and to prevent excessive pull on the tubing during installation. The tubing will regain its strength when its temperature returns to that of the surrounding soil, which usually occurs five minutes or less after installation.

The stiffness of tubing increases and its flexibility decreases as its temperature is lowered. Rapidly uncoiling tubing in cold weather stresses it excessively and may cause it to crack. The tubing may also have a tendency to coil in cold weather; it is then difficult to lay flat and must be handled with extra care. Ask the manufacturer for recommendations on handling the material in hot or cold weather. When plastic tubing floats in water during installation, it is difficult to get blinding material around and over the tubing without getting the material underneath it and causing misalignment. You can prevent floating by holding the tubing in place until blinding is completed.

Blinding Material

Blinding is the placement of bedding material consisting of loose, mellow soil on the sides and over the top of the drain to a depth of 6 inches. The bedding material must permit water to reach the drain easily. Except in areas where chemical deposits in and around the drain area are a problem, the bedding material should be friable top soil or other porous soil. Fine sand should not be placed directly on or around the drain. In soils with low permeability where blinding with soil is not adequate, a suitable envelope should be used. Blinding is not necessary where drains are placed in sand and gravel filters or envelopes.

A number of blinding methods have proven to be acceptable. Some contractors consider blinding so important that they place the material by hand around and over the drain. There are a number of mechanical blinding devices that can be mounted on the trencher. These devices take material from near the top of the trench and place it around and over the drain. Their main advantages are that they blind the drain immediately after laying it, use the most suitable blinding material that is readily available, and reduce labour requirements. All drains should be blinded immediately to maintain alignment and to protect them from falling rocks, ditch cave-ins, and backfill operations.Blinding immediately also will help maintain proper alignment of tubing in the groove and protect it when the remaining excavated material is placed in the trench.

Careful soil placement on both sides of the tubing is necessary to provide good side support, which will reduce deflection of the tubing. Hold tubing in place in the trench until it is secured by blinding. This step is especially important when water is in the trench and when the air temperature is below 45 *F*. Under those conditions, you may want to increase the quantity of blinding material. No stones or other hard objects should be allowed to come into contact with the drain. These objects apply point loads and may cause the drain to fail. Blinding provides protection for the drain during the backfilling operation when the impact of rocks and hard clods could damage it. All lines should be carefully inspected for grade alignment and other specifications before backfilling.

Backfilling Sediment

At the conclusion of each day's work, the end of the drain line should be stoppered and the trench backfilled to prevent sediment or debris from entering the line in case of rain. Backfilling should be done even sooner if there is a chance of heavy rain or freezing temperatures. The upper end of each drain should be tightly covered with a manufactured plug or equivalent material. Various methods can be used to move the remaining excavated material back into the trench and mound it up and over the trench to allow for settlement. These include graders, bulldozers, and auger and conveyer methods. The backfill material should be placed in the trench in such a manner that displacement of the drain will not occur. It is preferable to place the material on an angle so that it flows down the front slope. Avoid large stones, clods, and heavy direct loads during backfill operations. If you are installing tubing on a hot day and the tubing feels warm to the touch (1000 *F.*), delay the backfilling until the tubing reaches the soil temperature.

Trenchless Method of Installation

In the trench less plow method, tubing is placed at a prescribed depth in an open channel beneath a temporarily displaced wedge or column of soil. The trench less plow constructs a smooth-bottomed opening in the soil, maintains the opening until the tubing has been properly installed, and then surrounds it with permeable material.

The plow blade is designed to lift and split the overburden as it moves forward. The lifting action causes a deformation and disruption of the soil upward and outward at an angle on both sides of the plow blade. The slot should be fissured and loosened rather than compacted. The size of the shoe and drain-placing attachment should conform to the outside diameter of the tubing.

Critical Depth

The way the trench less plow moves through the soil influences whether the slot wall is fissured or compacted. A plow 3 to 5 inches wide working at a relatively shallow depth of 3 feet or less in dry soil will disturb the soil. The soil is broken loose from the base of the plow, heaves forward and upward ahead of the tine, and falls back around the tubing as the plow moves on. This type of disturbance creates cracks and fissures without causing compaction. The same plow working at a much greater depth of 6 to 7 feet in dry soil may tend to move the soil. Near the surface the soil is disturbed much as it is in drawing *A*. But below a certain depth, the plow may compress the soil sideways, causing it to become compacted and reducing its permeability. Between these two extremes of working depth there is a certain depth, termed critical depth, where the transition between one type of soil disturbance and the other occurs. This critical depth is the maximum depth at which the trench less plow can work without causing compaction around the tubing. The critical depth depends upon the forward inclination of the tine, the tine width, the soil moisture content, and the soil density.

In a uniformly compact, dry seed bed, the critical depth will be approximately 15 times the tine width. As the surface layers become very dry and strong or if the soil is of low bulk density and contains large air-filled pores, the critical depth will be reduced to about 10 times the tine width. You can sometimes lower the critical depth by loosening the surface layers of the soil to a depth of 8 to 16 inches in a strip approximately one and one-half times the working depth of the plow. The shallow tining can be done as the plow "returns empty" before the next run. An alternative is to use a double-pass system. The first pass with the plow is made at approximately two-thirds the final depth. Where the critical depth is above the working depth, using a wider tine and loosening the soil surface will reduce the draft force, often by a substantial amount.

Soil Condition of Texture and Moisture

Because soil texture and moisture vary considerably, the soil disruption patterns caused by the trench less plow will vary. The zone of disturbed soil extends upward and outward at various angles from the vertical, depending on soil conditions, depth, and plow geometry. Soils ranging from sands to clay-loam have properties that permit them to be fractured very readily by the plow.

In fact, when the surface layer of these soils is loose, it tends to flow down behind the tubing just as it would during a blinding operation. Wet soil that has high clay content is more likely than other soil types to develop soil structure and compaction problems. Both radial and surface compaction can occur where tubing is installed in heavy clay soils or wet soil. High-clay soils that are dry, however, can be fractured and fissured to a high degree, facilitating movement of water to the tubing.

Plows do not appear to be affected as much by rocks as are trenchers. In areas where rocks are a problem, the operator should flag any rocks he comes across so that they can later be inspected and removed if necessary. A plow will tend to push smaller rocks aside and move around the larger ones. When large rocks are encountered above or near grade, the plow tends to bounce over them, disrupting the tubing grade. If a rock deflects the plow upward a small distance, an alert operator will, where the slope permits, make a small adjustment in grade from that point on to avoid leaving a hump in the line that would have to be corrected later. If there is an adverse deviation in grade, the point of deviation should be marked. The tubing should be excavated at that point and the grade corrected. The grade should be controlled automatically by means of a laser.

Blinding Grade

To keep the tubing on grade, trench less plows should be fitted with a device that brings blinding soil over the tubing as it is placed. Where the surface layer is loose, it will flow down freely behind the tubing boot and cover the top of the tubing. This cover is desirable as long as the surface material is highly permeable and stable, but siltation could occur if the surface material is very fine. If a filter is used, a covering of medium to coarse sand around the tubing could enhance inflow.

Stringing Tubing Operate

Because trench less plows operate at high speed, it is sometimes difficult to keep pace when stringing the tubing in the field by hand before installation. To meet the requirements of high-speed plows and to reduce labour and materials costs, spools mounted on trailers have been developed.

Stretch Resistance

The amount of stretch depends on the stretch resistance of the tubing, the amount and duration of drag on the tubing as it is fed through the machine, and the temperature of the tubing at the time it is installed. Stretch may also be influenced by your method of handling the tubing when you string it in the field and the way the tubing is picked up and fed into the installation chute.

Tubing should not be stretched so much that its stiffness is reduced to less than the minimum allowable pipe stiffness. During installation, stretch should not exceed 5 per cent. We recommend that trench less plows be equipped with a power feeder to reduce stretch.

GENERAL CONSIDERATIONS OF DRAINS

Roads will affect the natural surface and subsurface drainage pattern of a watershed or individual hillslope. Road drainage design has as its basic objective the reduction and/or elimination of energy generated by flowing water.

The destructive power of flowing water, increases exponentially as its velocity increases. Therefore, water must not be allowed to develop sufficient volume or velocity so as to cause excessive wear along ditches, below culverts, or along exposed running surfaces, cuts, or fills.

Provision for adequate drainage is of paramount importance in road design and cannot be overemphasized. The presence of excess water or moisture within the roadway will adversely affect the engineering properties of the materials with which it was constructed. Cut or fill failures, road surface erosion, and weakened subgrades followed by a mass failure are all products of inadequate or poorly designed drainage. As has been stated previously, many drainage problems can be avoided in the location and design of the road: Drainage design is most appropriately included in alignment and gradient planning.

Hillslope geomorphology and hydrologic factors are important considerations in the location, design, and construction of a road. Slope morphology impacts road drainage and ultimately road stability. Important factors are slope shape (uniform, convex, concave), slope gradient, slope length, stream drainage characteristics (*e.g.*, braided, dendritic), depth to bedrock, bedrock characteristics (*e.g.*, fractured, hardness, bedding), and soil texture and permeability.

Slope shape gives an indication of surface and subsurface water concentration or dispersion. Convex slopes (*e.g.*, wide ridges) will tend to disperse water as it moves downhill. Straight slopes concentrate water on the lower slopes and contribute to the buildup of hydrostatic pressure. Concave slopes typically exhibit swales and draws. Water in these areas is concentrated at the lowest point on the slope and therefore represent the least desirable location for a road.

Hydrologic factors to consider in locating roads are number of stream crossings, side slope, and moisture regime. For example, at the lowest point on the slope, only one or two stream crossings may be required. Likewise, side slopes generally are not as steep, thereby reducing the amount of excavation.

However, side cast fills and drainage requirements will need careful attention since water collected from upper positions on the slope will concentrate in the lower positions. In general, roads built on the upper one-third of a slope have better soil moisture conditions and, therefore, tend to be more stable than roads built on lower positions on the slope.

Natural drainage characteristics of a hillslope, as a rule, should not be changed. For example, a drainage network will expand during a storm to include the smallest depression and draw in order to collect and transport runoff. Therefore, a culvert should be placed in each draw so as not to impede the natural disposition of stormflow. Culverts should be placed at grade and in line with the centerline of the channel. Failure to do this often results in excessive erosion of soils above and below the culvert. Also, debris cannot pass freely through the culvert causing plugging and oftentimes complete destruction of the road prism.

Headwater streams are of particular concern since it is common to perceive that measurable flows cannot be generated from the moisture collection area above the crossings. However, little or no drainage on road crossings in these areas is notorious for causing major slide and debris torrents, especially if they are located on convex slope breaks.

Increased risks of road failures are created at points *A* and *B*. At point *A*, water will pond above the road fill or flow downslope through the roadside ditch to point *B*. Ponding at A may cause weakening and/or erosion of the subgrade. If the culvert on Stream 1 plugs, water and debris will flow to point *A* and from *A* to *B*. Hence, the culvert at *B* is handling discharge from all three streams. If designed to minimum specifications, it is unlikely that either the ditch or the culvert at *B* will be able to efficiently discharge flow and debris from all three streams resulting in overflow and possible failure of the road at point *B*.

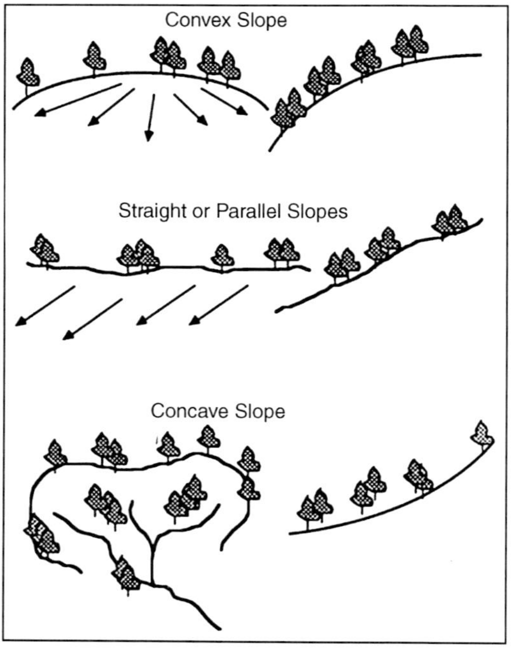

Fig. Slope Shape and its Impact on Slope Hydrology. Slope Shape Determines Whether Water is Dispersed or Concentrated. (US Forest Service, 1979).

A road drainage system must satisfy two main criteria if it is to be effective throughout its design life:

1. It must allow for a minimum of disturbance of the natural drainage pattern.
2. It must drain surface and subsurface water away from the roadway and dissipate it in a way that prevents excessive collection of water in unstable areas and subsequent downstream erosion.

The design of drainage structures is based on the sciences of hydrology and hydraulics-the former deals with the occurrence and form of water in the natural environment (precipitation, streamflow, soil moisture, etc.) while the latter deals with the engineering properties of fluids in motion.

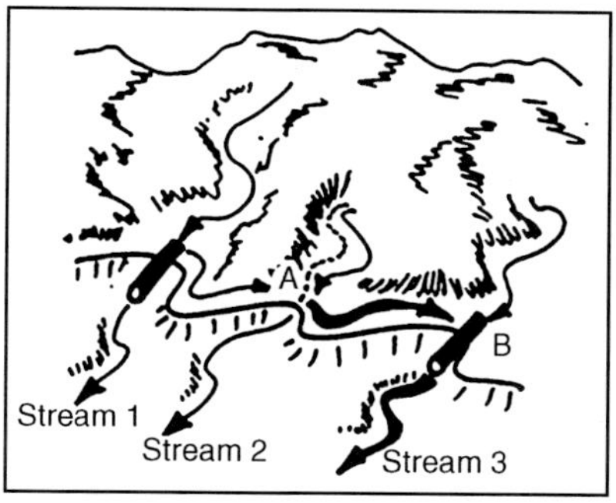

Fig. Culvert and Road Locations Have Modified Drainage Patterns of Ephemeral Streams 2 and 3. Locations *A* and *B* Become Potential Failure Sites. Stream 3 is Forced to Accept More Water Below *B* Due to Inadequate Drainage at *A*.

ESTIMATING RUNOFF OF DRAINAGE

Any drainage installation is sized according to the probability of occurrence of an expected peak discharge during the design life of the installation. This, of course, is related to the intensity and duration of rainfall events occurring not only in the direct vicinity of the structure, but also upstream of the structure. In snow zones, peak discharge may be the result of an intense warming period causing rapid melting of the snowpack.In addition to considering intensity and duration of a peak rainfall event, the frequency, or how often the design maximum may be expected to occur, is also a consideration and is most often based on the life of the road, traffic, and consequences of failure. Primary highways often incorporate frequency periods of 50 to 100 years, secondary roads 25 years, and low volume forest roads 10 to 25 years.

Of the water that reaches the ground in the form of rain, some will percolate into the soil to be stored until it is taken up by plants or transported through pores as subsurface flow, some will evaporate back into the atmosphere, and the rest will contribute to overland flow or runoff. Streamflow consists of stored soil moisture which is supplied to the stream at a more or less constant rate throughout the year in the form of subsurface or groundwater flow plus water which is contributed to the channel more rapidly as the drainage net expands into ephemeral channels to incorporate excess rainfall during a major storm event.

The proportion of rainfall that eventually becomes streamflow is dependent on the following factors:

- *The size of the drainage area.* The larger the area, the greater the volume of runoff. An estimate of basin area is needed in order to use runoff formulas and charts.
- *Topography.* Runoff volume generally increases with steepness of slope. Average slope, basin elevation, and aspect, although not often

called for in most runoff formulas and charts, may provide helpful clues in refining a design.

- *Soil*. Runoff varies with soil characteristics, particularly permeability and infiltration capacity. The infiltration rate of a dry soil, by nature of its intrinsic permeability, will steadily decrease with time as it becomes wetted, given a constant rainfall rate. If the rainfall rate is greater than the final infiltration rate of the soil (infiltration capacity), that quantity of water which cannot be absorbed is stored in depressions in the ground or runs off the surface. Any condition which adversely affects the infiltration characteristics of the soil will increase the amount of runoff. Such conditions may include hydrophobicity, compaction, and frozen earth.

A number of different methods are available to predict peak flows. Flood frequency analysis is the most accurate method employed when sufficient hydrologic data is available. For instance, the United States Geological Survey has published empirical equations providing estimates of peak discharges from streams in many parts of the United States based on regional data collected from "gaged" streams. In northwest Oregon, frequency analysis has revealed that discharge for the flow event having a 25-year recurrence interval. Most closely correlated with drainage area and precipitation intensity for the 2-year, 24-hour storm event. This is, by far, the best means of estimating peak flows on an ungaged stream since the recurrence interval associated with any given flow event can be identified and used for evaluating the probability of failure.

Table. Flood Recurrence Interval (Years) in Relation to Design Life and Probability of Failure.*

Design Life (years)	Chance of Failure (%)						
	10	20	30	40	50	60	70
	recurrence interval (years)						
5	48	23	15	10	8	6	5
10	95	45	29	20	15	11	9
15	100+	68	43	30	22	17	13
20	100+	90	57	40	229	22	17
25	200+	100+	71	49	37	28	21
30	200+	100+	85	59	44	33	25
40	300+	100+	100+	79	58	44	34
50	400+	200+	100+	98	73	55	42

* Based on formula $P = 1 - (1 - 1/T)n$, where n = design life (years), T = peak flow recurrence interval (years), P = chance of failure (%).

The probability of occurrence of peak flows exceeding the design capacity of a proposed stream crossing installation should be determined and used in the design procedure. To incorporate this information into the design, the risk of failure over the design life must be specified. By identifying an acceptable level of risk, the land manager is formally stating the desired level of success (or failure) to be achieved with road drainage structures. The flood recurrence intervals for installations in relation to their design life and probability of failure.

When streamflow records are not available, peak discharge can be estimated by the "rational" method or formula and is recommended for use on channels draining less than 80 hectares (200 acres):

$$Q = 0.278\, C\, i\, A$$

where:

Q = peak discharge, (m3/s)

i = rainfall intensity (mm/hr) for a critical time period

A = drainage area (km^2).

In English units the formula is expressed as: $Q = C\, i\, A$

where:

Q = peak discharge (ft^3/s)

i = rainfall intensity (in/hr) for a critical time period, tc

A = drainage area (acres).

The runoff coefficient, *C*, expresses the ratio of rate of runoff to rate of rainfall. The variable tc is the time of concentration of the watershed (hours).

Table. Values of Relative Imperviousness for Use in Rational Formula. (American Iron and Steel Institute, 1971).

Type of Surface	Factor C
Sandy soil, flat, 2%	0.05-0.10
Sandy soil, average, 2-7%	0.10-0.15
Sandy soil, steep, 7	0.15-0.20
Heavy soil, flat, 2%	0.13-0.22
Heavy soil, average, 2-7%	0.18-0.22
Heavy soil, steep, 7%	0.25-0.35
Asphaltic pavements	0.80-0.95
Concrete pavements	0.70-0.95
Gravel or macadam pavements	0.35-0.70

Numerous assumptions are necessary for use of the rational formula:

- The rate of runoff must equal the rate of supply (rainfall excess) if train is greater than or equal to tc;
- The maximum discharge occurs when the entire area is contributing runoff simultaneously;
- At equilibrium, the duration of rainfall at intensity I is t = tc;
- Rainfall is uniformly distributed over the basin;

- Recurrence interval of Q is the same as the frequency of occurrence of rainfall intensity I;
- The runoff coefficient is constant between storms and during a given storm and is determined solely by basin surface conditions.

The fact that climate and watershed response are variable and dynamic explain much of the error associated with the use of this method.

Manning's formula is perhaps the most widely used empirical equation for estimating discharge since it relies solely on channel characteristics that are easily measured. Manning's formula is:

$$Q = n{-}1\ A\ R2/3\ S1/2$$

where:

Q = discharge (m3/s)
A = cross sectional area of the stream (m^2)
R = hydraulic radius (m),
(area/wetted perimeter of the channel)
S = slope of the water surface
n = roughness coefficient of the channel.

In English units, Manning's equation is: $Q = 1.486\ n\text{-}1\ A\ R2/3\ S1/2$

where

Q = discharge (cfs)
A = cross sectional area of the stream (ft2)
R = hydraulic radius (ft)
S = slope of the water surface
n = roughness coefficient of the channel.)

Values for Manning's roughness coefficient are presented in Table below.

Table. Manning's *n* For Natural Stream Channels (Surface Width at Flood Stage Less than 30 *m*)

Natural stream channels	n
1. Fairly regular section:	
Some grass and weeds, little or no brush	0.030 - 0.035
Dense growth of weeds, depth of flow materially greater than weed height	0.035 - 0.050
Some weeds, light brush on banks	0.050 - 0.070
Some weeds, heavy brush on banks	0.060 - 0.080
Some weeds, dense willows on banks	0.010 - 0.020
For trees within channel, with branches submerged at high stage, increase above values by	0.010 - 0.020
2. Irregular sections, with pools, slight channel meander; increase values given above by	0.010 - 0.020
3. Mountain streams, no vegetation in channel, banks usually steep, trees and brush along banks submerged at high stage:	
Bottom of gravel, cobbles, and few boulders	0.040 - 0.050
Bottom of cobbles with large boulders	0.050 - 0.070

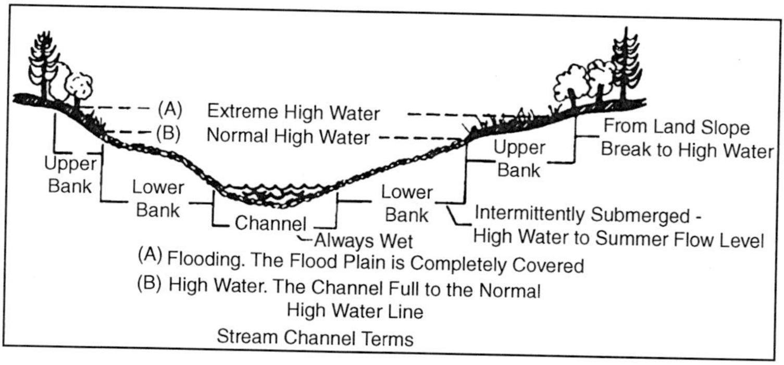

Fig. Determining High Water Levels for Measurement of Stream Channel Dimensions.

Area and wetted perimeter are determined in the field by observing high water marks on the adjacent stream banks. Look in the stream bed for scour effect and soil discoloration. Scour and soil erosion found outside the stream channel on the floodplains may be caused by the 10-year peak flood.

Examining tree trunks and brush in the channel and floodplain may reveal small floatable debris hung up in the vegetation. Log jams are also a good indication of flood marks because their age can be estimated and old, high log jams will show the high watermark on the logs.

The difficulty in associating high water marks with flow events of a specified recurrence interval makes values obtained by this method subject to gross inaccuracy. If the 10-year flood can be determined, flow levels for events with a higher recurrence.

Table. Relationship of Peak Flow with Different Return Periods.

Peak flow return period(years) (10-year peak flow = 1.00)	Factor of flood intensity
10	1.00
25	1.25
50	1.50
100	1.80

A key assumption in the use of Manning's equation is that uniform steady flow exists. It is doubtful that high gradient forested streams ever exhibit this condition. (Campbell, *et al.*, 1982). When sufficient hydrologic data is lacking, however, Manning's equation, together with observations of flow conditions in similar channels having flow and/or precipitation records, provide the best estimate of stream discharge for purposes of designing stream crossings.

An example illustrating the use of Manning's equation to calculate peak discharge is as follows:

- The wetted perimeter is equal to $10 + 2(5/\cos 45°) = 24.1\ m$.

- The cross sectional area is equal to 1/2 × sum of parallel sides x perpendicular height = 0.5(10 + 20)(5) = 75 m^2.
- The hydraulic radius is cross sectional area + wetted perimeter = 75/ 24.1 = 3.1 m.
- Manning's n from Table is n = 0.06
- Discharge, Q, from Manning's equation
  = (0.06)-1 (75)(3.1)2/3 (0.003)1/2 = 146 m^3/sec
  (Velocity, if needed, can be computed by Q/A = 1.9 m/sec.)

Location of Channel Crossings

Channel crossings require careful design and construction. Functionally, they must (1) allow for passage of the maximum amount of water which can reasonably be expected to occur within the lifetime of the structure and (2) not degrade water quality or endanger the structure itself or any downstream structures.

It should be pointed out that most road failures are related to inadequate water passage structures and fill design and placement as well as poor construction practices in such locations.

Accelerated erosion brought about by failure of channel crossing structures can be caused by:

- Inadequate design to handle peak flow and debris. Water will back up behind structure, saturating the fill and creating added hydrostatic pressure. Water will overflow the structure and the fil may be washed out.
- Inadequate outlet design. By constricting flow through a small area water velocity (along with its erosive power) will increase. Outlets need to be properly designed in order to withstand high flow velocities and thus avoid excessive downstream erosion and eventual road failure.
- Poor location of crossing. Crossings need to be located along relatively stable stretches where stream bottoms and banks exhibit little signs of excessive erosion or deposition. Meandering and/or multiple channels often indicate unstable conditions. If there is no choice but to use a poor location, careful consideration of the type of crossing selected, along with bank and stream bottom stabilization and protection measures, should be given.

There are three generally accepted methods used to cross channels on low volume roads--bridges, fords, and culverts.

The selection is based on traffic volume and characteristics, site condition (hydrologic/hydraulic conditions of channel), and management needs such as occasional closure, continuous use, safety considerations, resource impact (fish, wildlife, sediment).

Factors to consider when selecting a crossing type are listed as follows:

- *Bridges:* high traffic volume, large and variable water volume, high debris-potential, sensitive channel bottom and banks, significant fish resource, large elevation difference between channel and road grade
- *Culvert:* Medium to low water volume, medium to low debris potential, fish resource not significant, elevation difference between channel and road grade less than 10 meters, high traffic volume
- *Ford:* low to intermittent water flow, high debris potential, no fish resource, road grade can be brought down to channel bottom, low traffic volume.

All three channel crossing types require a careful analysis of both vertical and horizontal alignment. In particular, careful analysis of curve widening requirements is imperative in relation to the specified critical vehicle. Channel crossings are fixed structures where the road way width cannot be temporarily widened. Road width, curvature, approach, and exit tangents govern the vehicle dimensions which can pass the crossing. Except for bridge locations, roads should climb away from channel crossings in both directions wherever practical so high water will not flow along the road surface. This is particularly true for ford installations.

Fords

Fords are a convenient way to provide waterway crossing in areas subject to flash floods, seasonal high storm runoff peaks, or frequent heavy passage of debris or avalanches. Debris will simply wash over the road structure. After the incident, some clearing may be necessary to allow for vehicle passage.

Figure below shows a very simple ford construction where rock-filled gabions are used to provide a road bed through the stream channel.

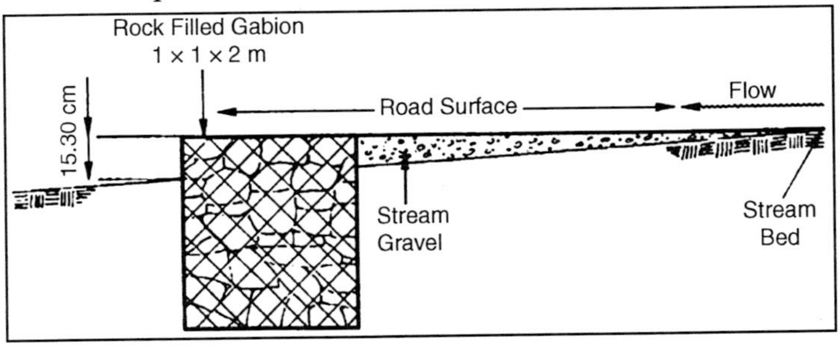

Fig. Ford Construction Stabilized by Gabions Placed on the Downstream end. (Megahan, 1977).

There are some design considerations which need careful attention:

- The ford should allow for passage of debris and water without diverting it onto the road surface. The ford results in a stream bed gradient reduction. Therefore, debris has a tendency to be deposited on top of a ford because of reduced flow velocity.

- Fords should be designed with steep, short banks which help to confine and channel the stream. The steepness and length of the adverse grade out of ford depends on the anticipated debris and water handling capacity required as well as vehicle geometry. Typically, the design vehicle should be able to pass the ford without difficulty. Critical vehicles (vehicles which have to use the road, but only very infrequently) may require a temporary fill to allow passage.

9

Subsurface Drainage System

Although subsurface drainage systems do not require extensive maintenance, the maintenance that is required is extremely important. If the subsurface drains are working, water will stand in the field for only a short time after a heavy rain. If water stands for a few days, the drain may be partly or completely blocked.

With drainage systems that have inspection wells or sediment traps, be sure to check the amount and rate of flow at these structures and at the outlets after a heavy rain. A change in flow may indicate that there is a blockage somewhere in the line. Regular inspection of the drainage system is essential. Prompt repair of any drain failure will keep the system in good working order and prevent permanent damage to it.

BENEFITS OF A SUBSURFACE DRAINAGE SYSTEM

Costing of drainage systems has shown that both increased pasture utilisation and extra pasture growth from drainage can be very competitive with other feed sources. However to take advantage of this extra growth, more cows, and farm improvements (including a feed pad) may be necessary. However, subsurface drainage provides many intangible benefits such as improvement in soil health due to the increased aeration of the soil, increased responses to fertiliser use, reduced mineral imbalances in the soil and sometimes in the plants as well, are long term sustainability benefits. Salinity can be prevented or greatly reduced if in its early stages.

Drained soils enables fodder conservation of silage/hay to occur several weeks earlier than on undrained soils. Therefore the resultant fodder has the potential to be of much higher quality producing improved animal production. Drainage, by reducing pugging and creating favourable soil environments, results in the retention of the improved pasture species versus the influx of plants suited to waterlogged conditions such as rushes, fog grass, glyceria species, water couch, etc. Most importantly, reduced stress in managing stock and pastures during the wetter months of winter or spring has very large benefits for the farm operator.

Types of Subsurface Drainage Systems

There are four main types of subsurface drainage systems.
These are:

- Corrugated and PVC slotted subsurface pipes
- Mole drainage including: Mole drains, Mole drains over collector pipe systems, Gravel mole drains
- Interceptor drains
- Ground water pumps

Subsurface Pipes

Subsurface pipes can be used to drain heavy (clay) poorly drained soils successfully but their spacing would need to be so close together that they are uneconomic in extensive farmland systems.

Fig. Drainage Trencher Installing Subsurface Pipe Drains and Depositing Permeable Backfill on Top

Subsurface pipe drainage was referred to as 'tile drainage' in the past due to the use of short clay pipes. Clay was expensive and difficult to lay and has now been replaced by slotted PVC or flexible corrugated plastic pipes of variable diameters.

Specifically designed drainage trenchers, usually fitted with laser guidance equipment, dig the trench, lay the slotted pipe and place permeable backfill into the trench on top of the laid pipe. This backfill is delivered by trucks or trailers fitted with conveyor belts which feed the backfill into the hopper. The forward speed, hopper channel opening size, material size, etc determine the depth and amount of material laid on top of the pipe. In very permeable soils, very little backfill is needed but in less permeable soils, or where moles are to be pulled through above a pipe, the backfill depth reaches to near the ground surface.

MOLE DRAINAGE

Mole drainage can be classified as mole drains, mole drains over a collector pipe system or gravel mole drains. The suitability of each type will depend on the clay content and type, sand and/or stone in the profile, gradient and outfall location. The action of the mole plough forms a mole channel in the area of the soil profile with a specific clay content. The plough also cracks the soil profile immediately above the mole channel allowing water to flow into it.

Mole Drains in Heavy Soils

Mole drains are used in heavy soils where clay subsoil near moling depth (400 to 600cm) prevents downward movement of ground water. The success and longevity of mole drains is dependent on soils having a high clay content so that once a mole channel is formed, it will maintain the channel for many years. Mole drains are not suited to soils with clay types which have dispersive or slaking characteristics. Mole drains are also not suited to permeable soils due to their high sand and/or loam contents.

A mole plough is used to form mole drains. Simply, a mole plough contains a leg (or blade) to which a torpedo (or foot) is attached to its bottom. Sometimes a plug (or expander) and having a slightly larger diameter, is attached to the rear of the torpedo, and ensures the mole channel is left with the correct shape.

Fig. Mole Plough

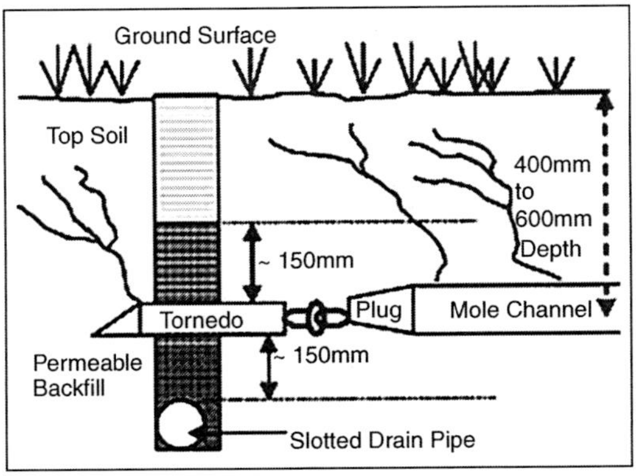

Fig. Mole Drain Over Collector Pipe System

Mole Drains Over Collector Pipe Systems

This system is used in soils where it is not possible to form moles that reach the outfall. This includes the presence of stones, sandy pockets, uneven surfaces or excessive distances to the outfall. In heavy soils where mole drains would need to be very long (over 80 *m*) before they reach an outfall, installing subsurface pipes at approximately 60 to 100 *m*, over which mole drains are pulled, can be very successful. Washed sand or small diameter gravel is backfilled into the pipe trench to near the ground surface at installation. Mole drains are then installed at or close to a right angle to the direction of the pipes. Excess ground water flows into and along the mole drains, then drains into the porous backfill above the pipes, and then is quickly removed to outfalls via the subsurface collector pipes.

Gravel Mole Drains

Gravel mole drains are best suited to soils and situations where subsurface pipes are unsuitable, where mole drains have a very short life span, or in slaking soils so the mole channel will maintain its shape at or soon after moling. A gravel mole drain is an unlined channel and/or leg slot filled with small diameter gravel or washed sand.

Fig. Gravel Mole Drain (Slot Filled with Gravel)

Fig. Gravel Mole Drain Machine

Unfortunately there are very few gravel mole drainage machines available in Australia. Also gravel mole drains are expensive due to the amount of backfill and the close spacing required. However, they do offer an alternative in some "difficult to drain" situations. They may be useful in slaking and dispersive soil types but expert opinion should be sought if considering their use in these situations.

Interceptor Drains System

These drains are installed at the base of slopes at the change of gradient, usually where a steeper slope meets the flats to intercept the downhill flow of subsurface water. Often the soil type on the slope is more permeable than those of the flats and this forces the water to come to the surface, usually at the change of slope.Interceptor drains can also be installed below springs and spring lines to intercept spring water. Grazing animals severely pug the areas surrounding springs and damage is usually more concentrated down slope. This affected area increases over time as the 'soak' area spreads outward and down slope. Drainage reduces stock damage, or pugging as the soil maintains its strength and so, structure.

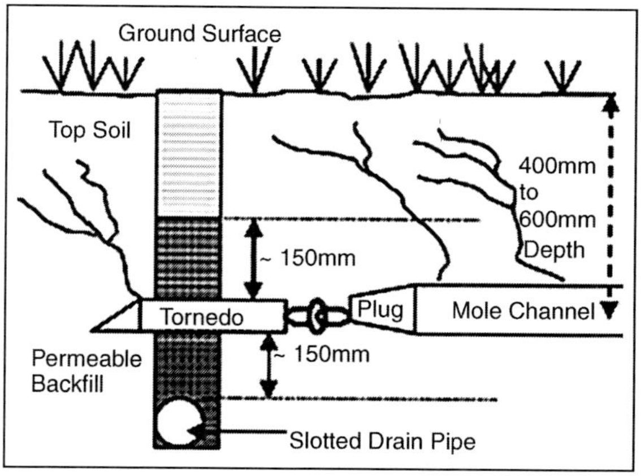

Fig. Interceptor Drain.

Ground Water Pumps System

These remove water from aquifers so that the water table will fall or be maintained at a suitable level below the ground surface. The act of pumping causes a drawdown of ground water leading out from the pump's location with its effect being much less at depth. The extent of effect will depend on aquifer depth, soil type, height of water table, etc.

The cost, benefits, disadvantages, and most importantly, whether they can be used or not and the need for a permit must be discussed with the Regional Rural Water authorities. Ground water pumping will not be discussed further in this Wet Soils Management series.

Assessing Which Drainage System to Install.

To decide which drainage system to install, the soil's characteristics, its permeability (speed at which water can move through the soil), and suitability for mole drainage (clay type and content) must first be determined. Farmers and drainage contractors can often decide which drainage option to use based on some simple on-farm tests. However, if there is any indecision from these on-farm tests, decisions should be backed up by soil tests and/or consultation with a subsurface drainage expert.

Water Act

The Water Act (1989) provides guidance for the management of waterways and swamps. Before considering draining a wet area you should contact your local Catchment Management Authority and Regional Water authority for advice, as a permit may be required.

TYPES OF DRAINAGE SYSTEMS

Agricultural land drainage usually consists of surface or subsurface systems, or a combination of both. At the field scale, subsurface drain pipes and field ditches normally exit to an open main or collector drain. At the regional level, the latter then empties into a river or its tributaries. In some instances, depending on the character of the hydrological basin, main drains may dispose of drainage water to an evaporation pond, to a wetland, or to a saline agriculture/ agriculture-forestry system. A schematic of drainage system components and several options for drainage water disposal within a watershed. Not all of these management options will necessarily be used in a single watershed.

Surface Drainage in Land Forming and Smoothing

Surface drainage is often achieved by land forming and smoothing to remove isolated depressions, or by constructing parallel ditches. Ditches and furrow bottoms are gently graded and discharge into main drains at the field boundary. Although the ditches or furrows are intended primarily to convey excess surface run-off, there is some seepage through the soil to the ditches, depending on the water table position. This could be regarded as a form of shallow subsurface drainage. Surface drainage is especially important in humid regions on flat lands with limited hydraulic gradients to nearby rivers or other disposal points. There is also a need for good surface drainage in semi-arid regions which are affected by monsoons.

Subsurface Drainage System

Surface drainage alone is seldom sufficient to remove excess water from the crop root zone. Deep ditches or subsurface pipe drainage systems enable a more rapid water table drawdown. The downstream ends of the laterals are normally connected to a collector drain. The required diameter of the pipe collectors increases with the area drained. Drain spacing is usually dependent on soil hydraulic conductivity and a design drainage rate coefficient. Depending on topography, land formation and proximity of a water receiving body, the collector may outlet by gravity to an open main drain or into a sump. In the latter case, the discharge is then pumped to another drain, or ultimately to a lake or stream.

In some flatter parts of eastern Canada, the eastern and mid-western United States, and parts of Europe, subsurface pipe drains are also used for sub-irrigation. In this case, in dry periods, surface water is introduced into the drain pipe system from an external source, and the water table is raised. Moisture then moves upward by capillary action to the root zone. Sub-irrigation is regarded as a highly energy and water efficient method of irrigation. In another process, known as controlled drainage, elevated water tables can be maintained with a control structure on the collector pipe.

Horizontal subsurface drainage systems are used in irrigated arid and semi-arid regions to reclaim saline and waterlogged lands, and to maintain favourable longterm salt and water balances in the crop root zone. Salinity and waterlogging are caused by a build up of the water table due to deep percolation of normal excess water and canal seepage. Buried pipe drains are generally installed deeper in arid regions than in humid regions in order to control salinity. Water in excess of plant evapotranspiration (ET) needs is always unavoidably applied during irrigation. This additional quantity of water applied is known as the leaching fraction. Naturally occurring as well as applied salts are then leached from the root zone by this water, and removed from the field via the pipe drains.

Deeper drain installation ensures that salts do not rise too rapidly to the soil surface due to capillary action. Drainage also prevents waterlogging of the root zone. The amount of irrigation water to be removed is generally less in arid than in humid regions. Vertical drainage by means of tubewells is also used to control waterlogging and salinity in some parts of the world, *e.g.*, India, Pakistan and central Asian republics. The primary purposes of tubewells are the same as those of horizontal drains, and at the same time to extract groundwater for irrigation. As a result of pumping, the water table is lowered, and salinisation due to capillarity is minimized. This situation is ideal where the groundwater is not very brackish or saline, and is therefore suitable for irrigation. In areas where the groundwater is highly saline, the pumped water may be too saline for irrigation, unless mixed with fresher or less saline water. Where the groundwater is too saline for crop production, it must be disposed of. Drainage does not have a direct impact on groundwater quality. It only serves to collect and transport excess water.

SECONDARY DRAINAGE TREATMENTS

Methods of improving the internal drainage of low permeability soils include: subsoiling, deep tillage, mole drainage, and biological practices, *viz.*, cropping with deep rooted legumes (*e.g.*, alfalfa) and crop rotations. In some parts of the world, deep rooted trees are used to lower the water table. There are usually no water quality hazards associated with these supplemental drainage practices.

Institutions of Environmental Impact Assessment

Many institutions require an environmental impact assessment (EIA) prior to construction activities associated with new projects or the rehabilitation of existing projects. The objectives are to identify potential adverse environmental effects, the magnitude of these effects, and to develop mitigative measures. Positive benefits are also identified. In cases where the adverse impacts far outweigh the expected benefits, the project may be completely redesigned or suspended. This is also the case where mitigative measures would be either too costly or technically not feasible. The EIA should be conducted in the earliest stages of decision making, when crucial decisions are still being deliberated.

Environmental effects can be classified as:

- *Direct and indirect:* Or first order and higher order. These are chain effects which are felt throughout, and possibly downstream of a catchment.
- *Secondary:* The primary activity of a drainage project may be extended to include secondary activities.
- *Synergistic:* These effects include an increased threat to the survival of certain species of wildlife that are under pressure in several ways as a result of the same project.

The ICID has developed an 'Environmental Checklist to Identify Environmental Impacts of Irrigation, Drainage and Flood Control Projects'. The World Bank has prepared an Environmental Assessment Sourcebook. FAO has produced a paper on the steps in the EIA process and the major environmental impacts of irrigation and drainage projects. The purpose of these documents is to enable detailed environmental impact assessments of irrigation and drainage projects to be conducted. The ICID checklist provides a comprehensive list of environmental parameters which must be evaluated in an EIA. Table shows the ICID checklist of possible environmental impacts of irrigation, drainage and flood control projects.

Ochs and Bishay listed the main steps in the EIA process as:

- *Scoping:* This is a public process, involving the participation of all parties. It results in specific guidelines for inclusion in the environmental impact study (EIS). (EIS refers to the document, whereas EIA refers to the entire process).
- *Drawing up the EIS:* Potential adverse and beneficial impacts are identified. Proposed actions with alternatives are indicated and mitigative measures are presented.

- Submitting the EIS for public review.
- Receiving advice from the reviewing agency.
- Accepting the EIS.
- *Choosing project components:* The decision-maker chooses the project option to be implemented and the mitigative measures.
- Implementing the project.
- *Monitoring:* During and after project implementation, the actual environmental impacts are monitored and compared with the EIS predictions. This is useful for improving future predictions and project designs.

Matrices and checklists are some of the tools most commonly used in an EIA to assess the magnitude or relative weight of both positive and adverse environmental effects. It is hoped that the criteria on drainage water quality provided in the publication will further assist in the evaluation of EIA of irrigation and drainage projects.

Water Quality Issues in Drainage Systems

The installation of drainage systems may result in changes to the associated ecosystem. These changes may be either beneficial or adverse. The positive environmental benefits were listed earlier in this chapter. However, there are potential adverse water quality impacts associated with drainage. The concentrations of salts, nutrients and other crop-related chemicals in drainage discharge vary with time and discharge rate. The use of fertilizers and pesticides in intensive agricultural production has sometimes led to damage to downstream aquatic ecosystems. Drainage planners therefore need to analyse effluent for nutrients and pesticides. The nutrients of most concern are N and P. In addition, from time to time, natural trace elements from the soil itself may be harmful to the ecosystem. Effluent laden with N and P stimulates eutrophication in receiving water bodies.

In addition to agricultural chemicals and trace elements, drainage water from irrigated areas frequently contains salts. The impact of salts on downstream users needs to be evaluated. Some soils are abundant in trace elements, and these could leach to the drainage system. Small amounts of trace elements such as As, Cd, Hg, Pb, B, Cr and Se are harmful to aquatic species because of biological magnification. The environmental consequences of disposing of drainage water from California's irrigated San Joaquin Valley into the 470 ha Kesterson Reservoir (a closed basin) are well known. The reservoir was a waterfowl habitat, and concentrating Se in the drainage water caused fish species to disappear, and resulted in deformities in waterbird embryos. The failure to construct an adequate outfall drain to the sea, the use of a waterfowl habitat for drainage water disposal, and the lack of proper water quality monitoring were the major reasons for these negative environmental impacts. Care must be taken to ensure that the disposal of drainage water does not interfere with the ecosystem's aquatic and terrestrial species. Concern has been expressed about damage by agricultural drainage water to estuarine fisheries in some countries.

Table. ICID checklist of possible environmental impacts of irrigation, drainage and flood control projects

Site:	Date:
Hydrology	1-1 Low flow regime
	1-2 Flood regime
	1-3 Operation of dams
	1-4 Fall of water table
	1-5 Rise of water table
Pollution	2-1 Solute dispersion
	2-2 Toxic substances
	2-3 Organic pollution
	2-4 Anaerobic effects
	2-5 Gas emissions
Soils	3-1 Soil salinity
	3-2 Soil properties
	3-3 Saline groundwater
	3-4 Saline drainage
	3-5 Saline intrusion
Sediments	4-1 Local erosion
	4-2 Hinterland effect
	4-3 River morphology
	4-4 Channel structures
	4-5 Sedimentation
	4-6 Estuary erosion
Ecology	5-1 Project lands
	5-2 Water bodies
	5-3 Surrounding area
	5-4 Valleys and shores
	5-5 Wetlands and plains
	5-6 Rare species
	5-7 Animal migration
	5-8 Natural industry
Socio-economic	6-1 Population change
	6-2 Income and amenity
	6-3 Human migration
	6-4 Resettlement
	6-5 Women's role
	6-6 Minority groups
	6-7 Sites of value

	6-8 Regional effects
	6-9 User involvement
	6-10 Recreation
Health	7-1 Water and sanitation
	7-2 Habitation
	7-3 Health services
	7-4 Nutrition
	7-5 Relocation effect
	7-6 Disease ecology
	7-7 Disease hosts
	7-8 Disease control
	7-9 Other hazards
Imbalances	8-1 Pests and weeds
	8-2 Animal diseases
	8-3 Aquatic weeds
	8-4 Structural damage
	8-5 Animal imbalances

ARTIFICIAL OF DRAINAGE SYSTEMS

A soil may need artificial drainage for either or both of the following two reasons:

- When there is a high water table that should be lowered; and
- When excess surface water cannot move downward through the soil or over the surface of the soil fast enough to prevent the plant roots from suffocation.

A holey soil seldom needs artificial drainage except when the soil has a high water table. If the problem exists at a low scale, any of the measures discussed above would suffice the need. Slowly permeable soils often need artificial drainage especially when the land surface is leveled and rainfall is high. Texture and structure affect the pore size distribution and drainage property of the soil. It is the predominance of small pore spaces which cause the restriction to drainage. The finer the texture, the smaller the size of pores and greater the need for artificial drainage. Conversely, the need for removal of excess water by artificial drainage is reduced as particle size increases from clay to silt, loam to sandy loam and sand. Soil structure *i.e* (the arrangement of soil particles) affects the drainage properties considerably. Soils having a platy structure have poor drainage properties as they resist the downward movement of water. Prismatic structures permits easy downward movement of water. Blocky structure offers favourable drainage characteristics.

Much can be learned about a soil's internal drainage by digging into the soil. Permeable soils that do not require drainage are uniform in colour throughout the profile. Yellow sub-soils usually indicate intermediate permeability, especially clay soils. Sub- soils which are mottled with red, yellow and grey are more slowly permeable than yellow sub-soils. Gray clay sub-soils in humid regions indicate very slow permeability and require drainage.

SURFACE DRAINAGE SYSTEM

Draining out surplus water from soil surface in order to provide favourable conditions for plant growth is called surface drainage. This type of drainage is mostly needed in plain lands with hard pan in the sub-soil.

Surveying and mapping of planned drainage area is necessary before starting any drainage programme. In These areas, topography should be clearly marked on the map where water accumulates. Surveying should be done just after raining. Information should be collected on areas where drainage water is to be disposed off, size of the drainage system, designed run-off, gravity flow, occurrence of flood, erosion and salt problems, etc. Surface drainage is also of two types as discussed below.

Relief Drainage System

This kind of surface drainage involves land forming like leveling, grading and construction of field ditches. Those ditches collecting drainage water of field are called collection ditches and the ones disposing off excess water are known as disposal ditches.

Leveling the Field Surface

By smoothening and leveling the field surface, water gets evenly spread over the whole surface. In such conditions, natural drainage *i.e* infiltration and percolation is accelerated. After leveling the field, a field drain is provided towards the lower end of the field. Water can be safely made to flow out of field through this drain.

Land Grading System

In this system, an uniform slope is provided in the field at the time of leveling. The slope percentage provided depends upon many factors like amount of rainfall, soil depth, soil texture, etc. If the field is extensive, it should be divided into graded segments or plots and each plot is provided with a field drain to dispose off water.

Surface Ditches Function

Two types of surface ditches are made according to their function.

Collection Ditches Function

According to their position, they are of the following types. Their function is to collect excess water of the field.

- *Field ditches:* They collect water within the field itself. Their cross section and size depends upon field topography, land area and land use pattern. They should be deep and wide enough to collect all the excess field water. Sides of the ditches should be plain.

- *Furrows:* For row crops grown on ridges, furrows virtually created between the rows, can be used for collecting excess water. These should be provided with slight slope towards main drainage channel.
- *Row ditches:* For such row grown crops which are very sensitive to water logging, small ditches can be created perpendicular to the crops row. These ditches meet with field ditch and they are deep as furrow.
- *Field drain:* This is a shallow graded channel with flat sides sloppy. They are created on the depressed side of the field and water from field furrows is delivered into them. To prevent soil erosion, grass cover should be allowed to grow on this drain.

Disposal Ditches Channels

These are lateral and main channels which are used for disposing off water of collection ditches. Their number is less than collection channels. The soil produced by digging the ditches should be put on the depressed spots of the field and used for leveling. This soil can also be used to provide a slight slope to the field.

Diversion Drainage Method

This is the method in which direction of the water flow is changed or directed towards a safer place and protected area is saved against possible damage of flooding. To provide diversion drainage, the structures like diversion ditches, contour ditches and flood way can be preventive measure for flooding.

Ditches can be dug by farmers with the help of spade or any other machine meant for this purpose. For leveling or digging ditches in extensive areas, use of machines is economically viable. Main machines used for these purposes are bulldozer, blade grader, dragline and claim shell. Latter one is used mostly if soil is permeable and loose.

Sub-surface Drainage

This kind of drainage is carried out by sub-surface drains which are artificial channels through which excess water may flow to a suitable outlet. The main purpose is to lower the ground water level below the root-zone of the plants. Sub-surface channels drain the sub-soil rather than surface soil, taking away only the excess water, not the water that plants can use. The excess water flows by gravity force into the drains and is carried through them to the outlet. Sub-surface drainage has the advantage over surface one. Subsurface drains occupy no land surface and they do not harbour weeds or interfere with farming operations. On most fields, however, a combination of sub-surface and surface drains provide the desired results. Subsurface drainage has the following specific advantages.

- Removes the gravitational or free water which is not directly available to plants and thus provides aeration and optimum soil temperature.
- Increases the volume of soil from which roots can obtain food.
- Permits the increase of bacterial activity in the soil which improves the soil structure and makes the plant food more readily available to the plants.
- Reduces soil erosion since a well drained soil has more capacity to hold rainfall resulting in less run-off.
- Improves soil moisture conditions in relation to the operation of tillage, planting and harvesting machines.
- Removes toxic substances such as alkali from the root zone.

Certain investigations are necessary before initiating any drainage measure. Information should be gathered on soil topography, soil characteristics, present land use pattern, crop rotation, quantity and distribution of rain, evapotranspiration, stream flow, subsoil horizons" hardpan, water table, etc. These information are the pail of survey and they should be properly mapped. Early development of sub-surface drainage began with narrow trenches in the field. Since these trenches hampered field operations, poles, brush stones or inverted wooden troughs were placed at the bottom of trenches which were then back filled. Usually these drains worked well for only a few years before they became obstructed by sediments or collapsed as the wood decayed. Modem practice uses tiles to provide sub-surface drainage. Mole drains are also used under favourable conditions. These are following important procedures of sub-surface drainage.

Pumping out Ground Water

The main objective of sub-surface drainage is to push down ground water level and stabilise it at a point where it is no more interfering with the root growth. In problem area, summer crops should be grown. Water should be pumped out from irrigation wells or ordinary bores to irrigate summer crops It serves two purposes; irrigation of crops and reduction of ground water level Analysis of this water for salt content should be done before using for irrigation purpose.

Mole Drains Dug

These drains are dug in the sub-soil with the help of mole plough. Water seeps into these drains and moves out to the main drainage channel. They were frequently used a few years back but now they are seldom used because of their short life. Properly installed mole drains can function for 3-4 years.

Use of Tile Drains

Tile drains are the most commonly used drains for underground drainage. They are very suitable for the areas where a great diversity or irregularity in topography and ground water levels occurs. Tiles are cheaper than other materials which could be used for underground drainage. They are not affected by the depth and spacing of tiles and depends upon the topography, soil permeability, *i.e.*, rate of infiltration and volume of excess water to be drained.

Practical limitations imposed by the depth of outlet available for the system and limitations of the trenching equipment have also to be considered. Depth and spacing of tile drains are usually determined by experience and judgement for a given set of conditions. To prevent breakage and shifting of the tiles, the minimum depth to the bottom of the tile is considered to be about 60 cm plus the tile diameter. In uniformally permeable soils the depth usually varies from 0.75 to 1.5 m depending on the soil type, depth and spacing and are usually kept as shown in the table given below.

Table. Average Depth and Spacing of tile Drains

Sl.No	Soil	Spacing (m)	Depth (m)
1	Clay	9-15	0.90-1.05
2	Clay loam	12-21	0.90-1.05
3	Loam	18-30	1.05-1.20
4	Sandy loam	30-60	1.20-1.50

Mixed Drainage System

In this system both the surface and sub-surface measures are employed to dispose off excess water. For soils which are very slowly permeable, buried tile drains alone may not be economically viable. In such lands, surface water is collected and diverted towards outlet by surface ditches and deep underground drains which are stabilised at wide spacing to drain out ground water.

BENEFIT OF DRAINAGE

Drainage provides a better environment for plant growth. The excess water which impedes the root respiration and which directly attacks the root tissues of most crops, is removed. The depth of the plant rooting zone is increased. Plants develop large root systems. The ability to grow an extensive root system means that the plant has a larger volume of soil from which it can extract nutrients and water.

Free water in the soil fills the pore spaces and air is excluded. The presence of air in the soil is essential for the growth of beneficial soil bacteria which converts the soil organic matter and fertilizer into available plant food. Drainage improves the soil structure and the infiltration capacity of soil is increased. Higher infiltration capacity reduces soil erosion. Optimum conditions for tillage are provided over a long range of time. inter-culture operation is made possible in the rainy season (kharif) crops. Less time and labour are required for tillage operation in a well drained soil. Crop damage at harvest time due to wet ground is provided.

Drainage hastens the warming of soils and maintain desirable soil temperatures. The removal of free water by drainage allows the soil to warm up more quickly because considerably less heat is required to raise the temperature of the well drained soil as compared to saturated soil. Proper soil temperature accelerates plant growth and bacterial activity. Conditions for early planting and better germination are thus improved by provision for drainage of soil.

Drainage promotes increased bleaching or salts and prevents their accumulation on the soil surface. Drainage not only improves the productivity of soil but also provides a more healthy environment for man and contributes to the general prosperity of a region.

OUTLET DITCHES IN DRAINAGE SYSTEMS

Many subsurface drainage systems fail because the outlet ditches are blocked. If the outlet ditch is filled with sediment, a survey should be conducted to determine how much cleanout work will be required. You should also find out whether some type of conservation practice could be used on the contributing watershed to reduce soil movement.

Surface Inlets

Poorly constructed surface inlets are subject to severe damage and require frequent repair Because inlet covers often become sealed with trash, they should be checked frequently. Clean the covers after a heavy rain and replace them carefully. If a cover is removed, trash can enter and block the line.

Blowouts Drains

Repair holes over subsurface drains at once. Otherwise, large amounts of soil may wash into the lines and block the entire system. Holes form where a drain is broken or where joints or slots are too wide. If the tile is broken, replace it. If the joint is too wide, place tile bats (pieces of broken tile) over the joint to prevent soil from washing into the line. To repair crushed or punctured corrugated plastic tubing, cut the damaged segment from the line and replace it with new tubing, using the manufacturer's couplers.

Sediment Traps in Drains

Sediment traps can be used for subsurface drains that are laid in fine sand or silty soils. If the traps are cleaned periodically, they will keep soil from filling the lines. Clean the traps every few days just after the lines are laid because at first sizeable quantities of fine soil will wash in through the joints between tiles or through perforations in plastic tubing. After one freezing and thawing cycle, soil will wash in more slowly, and you will need to check the traps only once or twice a year.

Tree Roots

Willow, elm, soft maple, cottonwood, and other water-loving trees that grow within approximately 100 feet of the drain should be removed. Maintain a clearance of 50 feet between the drain and other species of trees.

Ochre Accumulations

Ochre, which is an iron oxide, may block the drain when iron in solution moves from the soil to the drain and accumulates there. The process by which ochre accumulates may be chemical, microbial, or both. Ochre usually enters drains through organic soils but has been known to occur in other soils as well. There is no foolproof solution (except construction of open ditches) to the ochre problem. Jetting the drain with an acid solution has proven successful in some areas, but that remedy is very costly.

Pumping Plants

Where it is impossible or uneconomical to install outlets for drainage, drainage pumping plants can be used to remove excess surface water or groundwater. Pumping plants are also used where outlets are adequate except during prolonged periods of high water. If you are considering installing a pumping plant, be sure that its use is within the limitations ofIllinois drainage law. In solving drainage problems that involve pumping, take into account the capacity of the drainage system outlet, the capacity of the pump, its location and type, and the size of the sump. Determine the cost of all practical solutions. To be economically feasible, a pumping plant must be designed in such a way that annual operation costs are low. A pump that costs relatively little to install but that has a high annual cost of operation may not be the most economical.

A preliminary survey will determine the condition of the drainage outlet and help you decide whether pumping is required. A drainage system with a pumping plant that is designed into the system will usually function much more efficiently than one to which the pump is added later when the outlet is found to be inadequate.

The pumping plant must be designed to pump enough water to provide adequate drainage against the total head expected. (Total head considers all possible sources of resistance, from elevation to friction to couplings and joints.) Because pumping plants that are designed to pump surface runoff are complex and expensive, as much surface runoff as possible should be diverted from the site of the plant.

Selecting a Site of Pumping Plant

The pumping plant should be located where it can best serve its purpose. In choosing a location, consider the stability of the foundation material, accessibility for servicing, proximity to sources of power, and susceptibility to vandalism. In areas where ample sump storage is available, the pumping plant should be located so as to take maximum advantage of the storage. Select a location that will permit safe discharge into the outlet with a minimum of construction outside the diked area. If possible, locate the plant in a place that is readily accessible in all types of weather.

The requirement of a stable foundation is an important aspect in selecting a location. Before deciding upon a site, make borings to ensure that the location has the best foundation and that it meets as many of the other site requirements as possible.

Selecting Pumps Type

In selecting a pump, consider the type, characteristics, capacity, total head, the kind and source of power, shape and size of pump, housing, and method of operation. Pumps used for pump drainage are in the high-volume, low-head class. This class includes axial-flow propeller pumps and certain centrifugal pumps. For pumping from small capacity tile, you can use commercial sump pumps. Determine pumping volumes and heads carefully since friction factors become critical at settings other than those recommended by the manufacturer.

Electric power permits automatic operation and eliminates the need for daily fueling or servicing. Usually, a 10-horsepower motor is the largest that can be used on single-phase, 230-volt lines. Larger motors can be operated on three-phase power, which is available in some areas, or where phase converters can be used on single-phase power lines. If you plan to use electric power for pump drainage installations, consult the power supplier for suggestions and recommendations as to the best arrangement. If electric power is not available, you can operate the pump with diesel, gasoline, or *LP* gas stationary power units.

Belt or power takeoff drives can be used to couple farm tractors to the pump. The size of the pump depends upon the total head and the quantity of water pumped. The rated size is usually designated by the diameter of the pipe column at the discharge end of the pump. The design column velocity in the discharge pipe may range from 7 to 12 feet per second, with the highest efficiency usually occurring at values of 8 to 10 feet per second.

Pump Capacity of Plants

The required capacity of pumping plants can be determined from drainage coefficients applied to the area served, empirical formulas, a study of existing installations, or direct analysis using hydrologic procedures. The capacity of pumping plants for drainage areas of up to 1 square mile can be determined from applicable drainage coefficients or computed through hydrologic procedures. Hydrologic procedures should always be used whenever the drainage area exceeds 1 square mile. If you determine pump capacity from the drainage coefficient, use the following equation:

$Q = 18.9 \times C \times A$

where

- Q = pump capacity, gallons per minute (GPM),
- C = drainage coefficient (inches per 24 hours),
- A = area of the watershed (acres).

CONSISTS OF SURFACE DRAINAGE SYSTEM

A subsurface drainage system consists of a surface or subsurface outlet and subsurface main drains and laterals. Water is carried into the outlet by main drains, which receive water from the laterals. Submains are sometimes used off the main drain to collect water. The system will function only as well as its outlet. When planning a subsurface drainage system, make sure that a suitable surface or subsurface outlet is available or can be constructed. Where a surface outlet channel is used, all subsurface drains emptying into the outlet should be protected against erosion, against damage that occurs during periods of submergence, against damage caused by ice and floating debris, and against entry of rodents or other animals.

An older subsurface outlet used for a new subsurface drainage system should be free from breakdowns, fractured tile or crushed pipes, excessive sedimentation, and root clogging. It must be deep enough to intercept all outletting main drains and laterals and have sufficient capacity to handle the flow. It must also be deep enough to provide the minimum recommended cover for all drains newly installed or intercepted. If no suitable outlet is available and it is not practical to improve an existing ditch, you might consider using pump outlets.

RANDOM PATTERN

The random pattern is suitable for undulating or rolling land that contains isolated wet areas. The main drain is usually placed in the swales rather than in deep cuts through ridges. The laterals in this pattern are arranged according to the size of the isolated wet areas. Thus, the laterals may be arranged in a parallel or herringbone pattern or may be a single drain connected to a submain or the main drain.

Parallel Pattern

The parallel pattern consists of parallel lateral drains located perpendicular to the main drain. The laterals in the pattern may be spaced at any interval consistent with site conditions. This pattern is used on flat, regularly shaped fields and on uniform soil. Variations of this pattern are often combined with others.

Herringbone Pattern

The herringbone pattern consists of parallel laterals that enter the main at an angle, usually from both sides. The main is located on the major slope of the land, and the laterals are angled upstream on a grade. This pattern is often combined with others to drain small or irregular areas. Its disadvantages are that it may cause double drainage (since two field laterals intercept the main at the same point) and that it may cost more than other patterns because it contains more junctions. Nevertheless, the herringbone pattern can provide the extra drainage needed for the less permeable soils that are found in narrow depressions.

Double Main Pattern

The double main pattern is a modification of the parallel and herringbone patterns. It is applicable where a depression, frequently a watercourse, divides the field in which drains are to be installed. This pattern also is sometimes chosen where the depressional area is wet because of seepage coming from higher ground. Placing a main on each side of the depression serves two purposes: the main intercepts the seepage water, and it provides an outlet for the laterals. If the depression is deep and unusually wide and if you place only one main in the centre, you may have to make a break in the gradeline of each lateral before it reaches he main. By locating a main on each side of the depression, you an keep the gradeline of the laterals more uniform.

Site Considerations of Drainage

In planning a subsurface drainage system, the topography of the site to be drained, keeping in mind the depth limitations of the trenching machines and the amount of soil cover required over the drains. The amount of surveying you must do to obtain topographic information depends on the lay of the land. Where the slope of the land is obvious, only a limited amount of data is needed to locate the drains. A topographic survey is necessary, however, for flat and slightly undulating land since it is not as obvious where drains ought to be located. Obtain enough topographic information so that you can plan the entire system before installing it. Planning the job without first gathering enough data often results in a piecemeal system that may eventually be very costly.

The type of subsurface drainage system you install depends to a large degree upon the soils in the area to be drained. Knowing the soil types also helps you anticipate special drainage problems. To identify the soils, refer to soil maps that are available at local offices of the Natural Resources Conservation Service and the Cooperative Extension Service. Also in planning a subsurface system, keep in mind that trees such as willow, elm, soft maple, and cottonwood should be removed for a distance of approximately 100 feet on either side of a subsurface drain line. All other species of trees, except, possibly, fruit trees, should be removed for a distance of 50 feet. If the trees cannot be removed, plan to reroute the line or to use nonperforated tubing or tile with sealed joints throughout the root zone of the trees.

Patterns of Drainage

Because subsurface drainage is used primarily to lower the water table or remove excess water that is percolating through the soil over a general area, the drains are placed in a pattern determined by the characteristics of the area. If the soil is homogeneous, the water table is lowered at about the same rate on both sides of each drain. Flow from the drains is generally intermittent.

Four basic patterns are used in the design of subsurface drainage systems. Select the pattern that best fits the topography of the land, that can be located near enough to the sources of excess water, and that is suited to other field conditions.

PREPARATION AND DESIGNING DRAINAGE SYSTEMS

It is important to recognize that while technologies may be available to minimize water quality impacts from drainage effluent, institutional mechanisms must also be put in place to ensure that the technologies can be implemented. Policies and programmes, including legal and monitoring aspects, require an institutional framework.

It is recognized that, in some instances, there may be insufficient data in the project design and planning stages to enable precise conclusions to be drawn about possible environmental impacts. For this reason, most chapters contain sections on environmental monitoring, so that project planners and designers can elicit the information required for proper decision making. For more detailed information on water quality monitoring, the reader is referred to the United States Environmental Protection Agency (USEPA, 1982).

DISPOSAL AND ORGANIZATION OF DRAINAGE WATER IN CLOSED BASINS

Ultimate disposal of drainage water to a river or sea is not always possible. Closed drainage basins present a unique environmental or water quality challenge. In such situations, evaporation ponds may be an appropriate means for disposing of drainage water. However, these ponds may eventually lead to other environmental problems. For example, toxic substances could accumulate in the ponds. Furthermore, in arid climates, as pure water evaporates from the pond, the concentration of the remaining water approaches that of brine. The health of waterfowl, fish and other aquatic biota which use the pond could be negatively affected. Other environmental problems associated with evaporation ponds include: use of the pond to collect wastewater from homes; human health problems caused by consuming water from the ponds; and the need to ultimately dispose of the accumulated concentrated chemicals in the ponds.

In some cases, dry toxic materials may be spread by the wind. Furthermore, these ponds could become habitats for snails and mosquitoes, thereby causing malaria and schistosomiasis epidemics. In addition, if not properly managed, new waterlogged and saline areas will develop adjacent to the ponds. There is now increasing interest in the utilization of natural and constructed wetlands to manage drainage water. Wetlands are particularly effective for removing sediment, N and P. Plant, soil and hydrologic parameters interact in a complex way to filter and trap pollutants, and to recycle nutrients.

Certain tree and plant species have the potential to absorb pollutants. Residence time, flow rate, hydraulic roughness and wetland size and shape are some of the factors which influence treatment efficiency. The water supply to the wetland must be sufficient to provide an excess to discharge and prevent salt accumulation. In areas where soils, geologic and hydrologic conditions do not permit constructed wetlands, the saline agriculture/agriculture-forestry system may be appropriate for the disposal of drainage water. The operating principle is to successively re-use saline drainage water to irrigate crops and trees of increasing salt tolerance, and to discharge the final much reduced volume of water into a solar evaporator for salt crystallization. The depth of water ponded in the evaporator is regulated to match the daily evaporation rate.

The goal is to make the crystallization pond unattractive to flora and fauna. Another challenge of evaporation ponds is that evaporation is reduced as the ponds become more concentrated. This necessitates the use of additional land to maintain disposal (evaporation) capacity.

Water Stand Management

While there are instances where ecosystems have been damaged by poor quality drainage water, it is also possible to usa subsurface drainage systems beneficially to improve water quality. By managing the water table, through controlled drainage or sub-irrigation, nitrate concentrations in drainage effluent can be significantly reduced. Downstream flood flows can also be reduced. Water table management could therefore be viewed as a best management practice. Sub-irrigation and controlled drainage not only increase crop yields, but also enhance denitrification. Water table management is best suited to flat lands, and can be achieved by a simple modification to the outlet of subsurface drainage systems.

DRAINAGE USE FOR AGRICULTURE

Wetland soils may need drainage to be used for agriculture. In the northern USA and Europe, glaciation created numerous small lakes which gradually filled with humus to make marshes. Some of these were drained using open ditches and trenches to make mucklands, which are primarily used for high value crops such as vegetables. The largest project of this type in the world has been in process for centuries in the Netherlands. The area between Amsterdam, Haarlem and Leiden was, in prehistoric times swampland and small lakes. Turf cutting (Peat mining), subsidence and shoreline erosion gradually caused the formation of one large lake, the Haarlemmermeer, or lake of Haarlem. The invejtion of wind powered pumping engines in the 15th century permitted drainage of some of the marginal land, but the final drainage of the lake had to await the design of large, steam powered pumps and agreements between regional authorities.

The elimination of the lake occurred between 1849 and 1852, creating thousands of km^2 of new land. Coastal plains and river deltas may have seasonally or permanently high water tables and must have drainage improvements if they are to be used for agriculture. An example is the flatwoods citrus-growing region of Florida. After periods of high rainfall, drainage pumps are employed to prevent damage to the citrus groves from overly wet soils. Rice production requires complete control of water, as fields need to be flooded or drained at different stages of the crop cycle. The Netherlands has also led the way in this type of drainage, not only to drain lowland along the shore, but actually pushing back the sea until the original nation has been greatly enlarged. In moist climates, soils may be adequate for cropping with the exception that they become waterlogged for brief periods each year, from snow melt or from heavy rains. Soils that are predominantly clay will pass water very slowly downward, meanwhile plant roots suffocate because the excessive water around the roots eliminates air movement through the soil.

Other soils may have an impervious layer of mineralized soil, called a hardpan or relatively impervious rock layers may underlie shallow soils. Drainage is especially important in tree fruit production. Soils that are otherwise excellent may be waterlogged for a week of the year, which is sufficient to kill fruit trees and cost the productivity of the land until replacements can be established. In each of these cases appropriate drainage carries off temporary flushes of water to prevent damage to annual or perennial crops.

Drier areas are often farmed by irrigation, and one would not consider drainage necessary. However, irrigation water always contains minerals and salts, which can be concentrated to toxic levels by evapotranspiration. Irrigated land may need periodic flushes with excessive irrigation water and drainage to control soil salinity.

SUBSURFACE DRAINAGE

Subsurface drainage is the removal of water from the rootzone. It is accomplished by deep open drains or buried pipe drains.

Deep Open Drains

The excess water from the rootzone flows into the open drains. The disadvantage of this type of subsurface drainage is that it makes the use of machinery difficult.

Pipe Drains

Pipe drains are buried pipes with openings through which the soil water can enter. The pipes convey the water to a collector drain.

Drain pipes are made of clay, concrete or plastic. They are usually placed in trenches by machines. In clay and concrete pipes (usually 30 cm long and 5 - 10 cm in diameter) drainage water enters the pipes through the joints. Flexible plastic drains are much longer (up to 200 m) and the water enters through perforations distributed over the entire length of the pipe.

Deep open drains versus pipe drains

Open drains use land that otherwise could be used for crops. They restrict the use of machines. They also require a large number of bridges and culverts for road crossings and access to the fields. Open drains require frequent maintenance (weed control, repairs, etc.).

In contrast to open drains, buried pipes cause no loss of cultivable land and maintenance requirements are very limited. The installation costs, however, of pipe drains may be higher due to the materials, the equipment and the skilled manpower involved.

SURFACE-WATER INLETS AND SUBSURFACE DRAINS

Surface-water inlets allow surface water to enter subsurface drains. Because of the high cost of carrying surface water in buried drains, inlets are recommended only for draining low areas where it is not feasible to install a surface drainage system. If you have to use surface-water inlets, place nonperforated tubing or conduit on each side of the riser. Since surface-water inlets may be a source of weakness in a drainage system, you might consider offsetting the inlet to one side of the line to reduce the hazard to the main line. Surface-water inlets not projecting above the surface should be protected with a cone grate (often referred to as a "beehive").

The cone grate tends to float flood debris, preventing it from closing off the entrance. Metal pipes and other durable materials with holes or slots may be used as inlets. Flow control devices may be necessary to limit the amount of water entering the drain. One way of limiting water flow is to install an orifice plate near the bottom of the inlet. Where it is likely that a substantial amount of sediment will enter the surface-water inlet, it is advisable to construct a sediment trap.

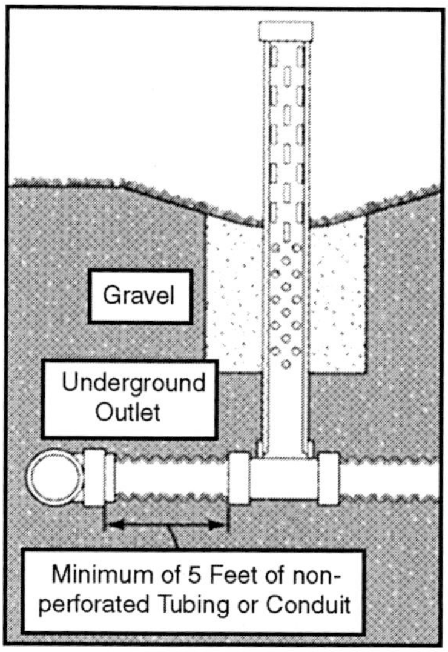

Fig. A Surface-Water Inlet Offset from the Subsurface Drain Line to Reduce Potential Damage to the Line.

Filters and Envelopes of Drain

The need for a filter or envelope depends upon the characteristics of the soil material at drain depth and the velocity of flow in the conduit. Filters may be required in sand, silt, and some organic soils to prevent sediment from accumulating in the drain. A filter is required where the base material is uniform, fine to medium sand and where flow reaches such high velocities that it moves the sand into the drain.

Filters may be sand and gravel envelopes or manufactured filter material. Most of the presently available, artificial, prefabricated filter materials, such as fiberglass, spun-bonded or knitted synthetic fabrics, and plastic filter cloth, act as protective filters. With time, however, these materials may partially clog and decrease flow into the drain. Manufactured filters should have openings of sufficient size and enough strength, durability, and permeability to provide constant filtering of the soil material and to protect the drain throughout the expected life of the system. Make sure that the manufacturer of the material has certified it for underground use. During installation the material should span all open joints and perforations without being stretched excessively. Be careful not to damage the material during installation. Any damaged areas should be replaced before backfilling. Installing envelope material around subsurface drains provides them proper bedding support and improves the flow of groundwater into the drain. Where it is not feasible to form a bedding groove for plastic drain tubing, envelope material can be substituted for bedding. The minimum thickness of the envelope may vary from 3 to 6 inches, depending on the type of equipment used to install the material and the availability of the gravel material. For all envelope designs, if the trench is wider than the specified width of the envelope, the trench must be filled on both sides with bedding material or a gravel envelope so that no space is left between the drain and the walls of the trench.

10

Stages of Land Preparation for Farming

Land preparation is a key aspect of farming, as it promotes the rapid emergence of crop seedlings and provides ventilation, water and nutrients for crop growth. Preparation stages vary according to the soil type and crops being planted. Improper land preparation can result in soil erosion, crusting weed overgrowth or water-logging.

THE IMPORTANCE OF LAND PREPARATIONS

Levelling, smoothing and shaping the field surface is as important to the surface system as the design of laterals, manifolds, risers and outlets is for sprinkler or trickle irrigation systems. It is a process for ensuring that the depths and discharge variations over the field are relatively uniform and, as a result, that water distributions in the root zone are also uniform. These field operations are required nearly every cropping season, particularly where substantial cultivation following harvest disrupts the field surface. The preparation of the field surface for conveyance and distribution of irrigation water is as important to efficient surface irrigation as any other single management practice the farmer employs.

There are perhaps two land levelling philosophies:

1. To provide a slope which fits a water supply; and
2. To level the field to its best condition with minimal earth movement and then vary the water supply for the field condition.

The second philosophy is generally the most feasible. Because land levelling is expensive and large earth movements may leave significant areas of the field without fertile topsoil, this second philosophy is also generally the most economic approach.

Land levelling always improves the efficiency of water, labour and energy resources utilization. The levelling operation, however, can be the most intensively disruptive cultural practice applied to the field and several factors should be considered before implementing a land levelling project. Major topographical changes will nearly always reduce crop production in the cut areas until fertility can be replaced. Similarly, equipment traffic can so compact or pulverize the soil that water penetration is a major problem for some time. The farmer has many activities which contribute to his productivity and therefore require his skill and labour. The irrigation system should be designed with him (or her) in mind. A field levelled to high standards is generally more easily irrigated than one where undulations require special attention.

New equipment is continually being introduced which provides the capability for more precise land levelling operations. One of the most significant advances has been the adaptation of laser control in land levelling equipment. The equipment has made level basin irrigation particularly attractive since the final field grade can be very precise. Comparisons with less precise techniques have clearly shown that laser–levelled fields achieve better irrigation and production performance. Nevertheless, for most irrigated agriculture, laser–controlled precision is unfeasible because of the high cost of such equipment unless a large number of farmers form a cooperative or a government programme is started with subsidized land levelling as one component in an effort to improve farm production.

SMALL-SCALE LAND LEVELLING

Most small–scale farming operations rely on animal power or small mechanized equipment which an individual can own and operate. As the irrigator waters his fields season after season he is able to observe the locations of high and low spots on the field. Then as he prepares the fields between plantings, he tries to move soil from the high spots to the low ones. Over a period of several years individual fields are smoothed enough to be watered fairly well. In a farmer is preparing land for paddy and using the ponded water level on the field to direct him to the high and low spots. Since this is a normal land preparation practice, it does not represent an extra task for the irrigator. A similar operation using mechanized equipment for typical annual crops and again one sees that the field preparation also readies the seed bed for planting. Beyond these technologies one may observe various levels of mechanization and an array of implements. The one feature common to most small–scale land levelling is the trial and error nature of the practices and the long-term incorporation of land levelling with seed bed preparation. Another feature is that no technical or engineering inputs are needed.

Importance

- Preparing farmland before planting places seeds in closer contact with soil moisture, allowing for more effective germination. Tillage overturns soil to prevent crusting and allows air and water to penetrate. It also removes weeds, which compete with crops for light and nutrients. Irrigation provides soil with water crop growth. Fertilizers and pesticides are added to soil to provide essential crop nutrients and limit crop damage by pest infestation.

Tillage

- Tillage involves overturning of topsoil. Depending on the scale of farming, farmers till using simple tools, such as a hoe or shovel, or complex equipment such as tractors. Tillage is divided into three categories: conventional, reduced and conservation tillage. Conventional tillage is uncommon as it contributes extensively to soil erosion. Reduced tillage leaves between 15 to 30 per cent crop residue on the soil. Conservation tillage leaves more than 30 per cent residue, and sometimes includes no tillage at all.

Irrigation

- Irrigation systems are designed according to water availability, quantity, quality and the distance between the water source and farmland. A system consists of a pumping station and conveyance, distribution, field application and drainage systems. The pumping station channels water from a source such as a river into the irrigation system. The conveyance and distribution systems ensure water reaches from the pumping station to the farmland. The field application system transports water throughout the farmland, and the drainage system eliminates excess water.

Fertilizers

- Fertilizers are either organic, such as manure and crop residue, or chemical. Fertilizing soil provides important nutrients such as nitrogen, phosphorous, potassium, calcium, magnesium and sulfur. Regardless of the type, fertilizers add ammonium to soil, which increases its acidity. Lime is added to soil to combat this effect, a reaction which produces water and carbon dioxide.

LAND PREPARATION

Land preparation typically involves plowing, harrowing, and leveling the field to make it suitable for crop establishment. Draft animals, such as buffalo and oxen, 2–wheel tractors or 4–wheel tractors can all be used as power sources in land preparation. The initial soil tillage can also be performed with a rotovator instead of a plow.

Flooded Soils

The vast majority of Asian rice fields are first flooded with water before tillage. This tillage of flooded soil is referred to as puddling. Soil flooding and soaking is performed once and requires sufficient water to bring the topsoil to saturation and create an overlying water layer. Soil puddling destroys soil structure, which reduces percolation rates and loss of water. It also results in high resistance to root penetration, low porosity, and permeability and in the formation of a soil plow pan; all of which can restrict root growth. Puddling is very efficient in clay soils that form deep cracks penetrating the plow pan at about 15 to 20 cm soil depth during the period of soil drying before land preparation. Although puddling reduces percolation rates of the soil, the action of puddling itself consumes water. There is a trade-off between the amount of water used for puddling and the amount of water "saved" during the crop growth period as a result of reduced percolation rates. Puddling is less effective in coarse soils, which do not have enough fine clay particles to migrate downward and fill up the cracks and pores in the plow pan.

Non Flooded Soil

A well-levelled field is a prerequisite for good water and crop management. When field are not level, water may stagnate in the depressions whereas higher parts may fall dry. This results in uneven crop emergence and uneven early growth, uneven fertilizer distribution, and possibly additional weeds.

Effective land levelling will:

- Improve crop establishment and care,
- Reduce the time and water required to irrigate the field, and
- Ensure more uniform distribution of water in the field

Land preparation for dry seeding typically involves consists of plowing or rotovating followed by harrowing and levelling of dry and friable soil. The crop is established by broadcasting or drilling seed that has not been pre germinated.

Farmers can reduce water use by shifting from puddled to non-puddled land preparation. Large amounts of water (20–40 per cent of total water use) are consumed during land preparation of flooded soils because of the need to initially soak dry and cracked soil, and to keep the field continuously flooded. Most of the wasted water is lost by drainage through soil cracks. Dry land preparation does not require irrigation water because it can be done when the soil has the correct water contact and is friable for plowing or rotovating and harrowing.

AGRARIAN STRUCTURE LAND TENURE, OWNERSHIP, TENANCY AND CULTIVATION

The interim report the Subcommittee on Land Policy, Agricultural Labour and Agricultural Insurance was considered by the National Planning Committee of the Indian National Congress at its fourth session at the end of June, 1940. The Committee resolved that "the cooperative principle should be applied to the exploitation of land by developing collective and cooperative farms... collective or cooperative farms should be begun on 'cultivatable waste land, which should be acquired, where necessary by the state immediately". Professor Radhakamal Mukherjee, a member of the committee did not want to rule out peasant farming in small heritable holdings and wanted it to continue along with cooperative and collective farming. At its fifth session in September, 1940 the chairman of the Land Policy Subcommittee presented a further note. After consideration of the note, the full committee reiterated its earlier resolution that "cultivation of land should be organised in complete collectives; wherever feasible... other forms of cooperative farming shall be encouraged elsewhere."

After independence, the committee on agrarian reforms in its report of 1950, concluded that collective farming to be suitable essentially for the development of reclaimed waste land. After examining three other alternatives, namely capitalist farming, state farming and individual peasant farming, it categorically rejected capitalist farming as its adoption in its view "would deprive the agriculturists of their rights in land and turn them into mere wage earners," was unenthusiastic about state farming, except once again, on reclaimed waste land, and opted for individual peasant farming.

The idea of cooperative farming surfaced in the form "cooperative village management" in the First Five Year plan with the village as the unit of land management with individual families or groups of families cultivating blocks of land allotted by the village management body. However, right of ownership of the village land would be recognised and compensated through an ownership dividend at the end of each harvest. Dandekar acidly comments that "This was a rather naive concept based on a utopian notion of a village and plain ignorance, or unwillingness to see the truth, about village community functioned."

The Second plan, according to Dandekar "offered lip service though with less conviction," to cooperative village management and the third plan made no mention of village management and thereafter the concept was quietly dropped. The idea of bringing together holdings below a certain level into small cooperative farms did not proceed very far either. It also went out of consideration after the Third plan.

Although the problem of landless agricultural workers was recognised and the need to provide increased employment opportunities (on and off farm) was also recognised as Dandekar points ideas on increasing employment opportunities "were not very clear, in any case, they were not elaborated... what was said with respect to the landless workers in the First Five Year Plan was plainly evasive." In particular, there was no understanding that the development strategy being capital intensive by its very design could not generate the rising employment opportunities for such workers. Indeed the implicit presumption then was the problem of their employment was to be solved within the agricultural sector itself. Dandekar points out that "the Eighth Plan emphasised that landlessness was a root cause of poverty and that access to land was a major source of employment and income; that such access could be achieved either by a more equitable distribution of land or providing security of tenure to tenants and share croppers who are the actual cultivators."

In his summing up of the official approach to transforming the traditional agriculture, Dandekar correctly argues that the so-called "land problem," which the First Five Year Plan claimed overshadowed all other problems, was "an excessive burden of population which the land has to bear and a satisfactory solution is supposed to be to let the land continue to bear this burden," again illustrating the lack of recognition of the lessons of development history. With land continuing to bear the burden, the land policy was land reform: ceilings on land holdings and security of tenure for tenants. Again Dandekar is right in stressing that the land reform policy did not recognise that by keeping a growing rural population on land, land reforms simply created a growing population of non-viable farmers (small and marginal farmers). The history of legislation on land ownership, illustrates Dandekar's point.

LAND POLICY

Land policy in India has been a major topic of government policy discussions since the time prior to Independence from British rule. The peasants of the country strongly backed the independence movement and the "Land to the Tiller" policy of the Congress Party because of the prevailing agrarian conditions. The agrarian structure during British administration emerged with a strong historical background. The land-revenue system implemented by Todar Mal during Akbar's regime can be traced as the possible beginning of systematic efforts to manage the land. This method incorporated measurement, classification and fixation of rent as its main components.

Under the various pre- British regimes, land revenues collected by the state confirmed its right to land produce, and that it was the sole owner of the land. British rulers took a cue from this system and allowed the existence of noncultivating intermediaries. The existence of these parasitic intermediaries served as an economic instrument to extract high revenues (Dutt, 1947) as well as sustaining the political hold on the country. Thus at the time of Independence the agrarian structure was characterized by parasitic, rent-seeking intermediaries, different land revenue and ownership systems across regions, small numbers of land holders holding a large share of the land, a high density of tenant cultivators, many of whom had insecure tenancy, and exploitative production relations.

Immediately after Independence a Committee, under the Chairmanship of the late Shri J. C. Kumarappa (a senior Congress leader), was appointed to look into the problem of land. The Kumarappa Committee's report recommended comprehensive agrarian reform measures. India's land policy in the decades immediately following its independence was dominated by legislative efforts to address the problems identified by the Kumarappa Committee. A substantial volume of legislation was adopted, much of it flawed and little of it seriously implemented.

Several important issues confronted the policy-makers.

1. Land was concentrated in the hands of a few and there was a proliferation of intermediaries who had no vested interest in self-cultivation. Leasing out land was a common practice.
2. The tenancy contracts were expropriative in nature and tenant exploitation was ubiquitous.
3. Land records were in extremely bad shape giving rise to a mass of litigation. It is ironic that the Supreme Court of India in 1989 commented that the revenue records are not legal documents of title. This is a sad commentary on the land records of the country.

It is against this background that land policy has been shaped in India. While land-reform legislation remained active, land policies in more recent decades have focused less on land reform and more on land development and administration.

Land policy in India has undergone broadly four phases since Independence.

1. The first and longest phase (1950 - 72) consisted of land reforms that included three major efforts: abolition of the intermediaries, tenancy reform, and the redistribution of land using land ceilings. The abolition of intermediaries was relatively successful, but tenancy reform and land ceilings met with less success.
2. The second phase (1972 - 85) shifted attention to bringing uncultivated land under cultivation.
3. The third phase (1985 - 95) increased attention towards water and soil conservation through the Watershed Development, Drought-Prone Area Development (DPAP) and Desert-Area Development Programmes (DADP). A central government Waste land Development Agency was established to focus on wasteland and degraded land. Some of the land policy from this phase continued beyond its final year.
4. The fourth and current phase of policy (1995 onwards) centres on debates about the necessity to continue with land legislation and efforts to improve land revenue administration and, in particular, clarity in land records.

This chapter is an attempt to discuss the critical issues pertaining to land policy in India beginning with the emergence of a post-Independence policy in a historical context, and from a viewpoint of differential provincial policies. Naturally, land reforms predominate. Land reforms have been one of the important land policy initiatives in India that have brought a fundamental change in the entire approach towards development. The impact of land reforms and the changing phases of land administration. The focus is on the role and development process of land policy in India in the context of overall changes in India's development policies.

It also addresses the political and economic aspects of the policy initiatives, beginning with the various land-reform efforts and finally analysing the recent land development and administration policies. The paper incorporates a discussion on the closely related goals of land policy, *i.e.* poverty elimination, conflict management, sustainable economic growth and good environmental management. In the final analysis, the paper highlights current issues pertaining to the relationships among land policy, poverty and the development initiatives.

EMERGENCE OF LAND POLICY IN INDIA

Given India's vastness, diversity and various political, economic and social influences from a history of various rulers and foreign conquerors, it is not surprising that land tenurial and administration practices varied significantly throughout the subcontinent at the time of Independence. One common factor was that land policies had been driven by the rulers' efforts to extract land revenue or tax from those working on the land. Throughout much of the country, the rulers appointed Zamindars, or tax collectors, who were contracted to collect land revenue for a given large territory and pay fixed sums to the government (but often extracted as much as they could from the landholders and pocketed the difference).

Though tenurial conditions varied significantly from region to region, the numerous tenures could be classified under two broad categories - the Zamindari and Rayatwari systems. The Zamindari system was characterized by one or more layers of proprietary rights between the state and the actual landholder. In the Rayatwari (or peasant proprietorship) system, no intermediaries existed in design but emerged in the process. The British rulers continued with existing land-revenue policies and procedures with a few but significant modifications. Perhaps most importantly, the British made the tax-collecting Zamindars into proprietors of the estates over which they had tax collection duties.

This change was aimed at accomplishing two objectives: simplifying the land-revenue collection process and creating a rural elite with a vested interest in British rule. Unfortunately, it converted the erstwhile landowners into insecure tenants. Over time, many Zamindars assigned their land-revenue collection duties to one or more layers of intermediaries who were also given interests in the land. The historical emergence and perpetuation of intermediaries served the purpose of land revenue administration and political control of the successive rulers, but their numbers swelled. The large patches of land held by them were let to tenants at exorbitantly high rents. That created a disincentive among the tenant cultivators to develop the land, and consequently impacted upon production. Thus, the Colonial Government, out of its interest to administer the country effectively, did not make any substantial changes in the land-revenue system but promoted the class of non-cultivating intermediaries.

At the time of Independence, India faced a major challenge of setting right the agrarian structure as promised during the independence struggle. Thorner and Thorner (1961), in an analysis of the agrarian structure of India, vividly describe the pre-Independence structure as a complex of legal, economic and social relations - a multilayered structure that pulled down the production efficiency in the agricultural sector.

A brief review of the literature also reveals a myriad of agrarian relations in India, varying from peasant proprietorship to a pure landlord - serf relationship. The first task placed before the first Indian parliament was to address land policy. Because India has a densely populated agrarian economy, almost all other developmental initiatives also involved land as a central and a complex issue, as it clearly represented social status and not just the means of production.

While recognizing the need to bring about land reforms in the country, the Constitution of India provided under Article 39 that: (1) the ownership and control of the material resources of the country should be so distributed as best to serve the common good; and (2) the operation of the economic system should not result in a concentration of wealth or a means to production to the common detriment.

The Constitution of India also made land a state (provincial) subject. So, only state (provincial) legislatures have the power to enact and implement land-reform laws. However, the central government played a significant advisory and financial role in land policy based on its constitutional role in social and economic planning (a role held concurrently with the states). The Government of India established a National Planning Commission immediately after Independence to fulfil this role of social and economic planning.

The Planning Commission has prepared a series of Five-Year Plans since 1951. Land policy has been one of the important components incorporated in all the plans. The policy statements are sometimes quite explicit in the plan documents, but are more often implicitly stated. An overview of changes in the land policy as reflected through the various plan documents. Land reform policy was spelt out in the First Five-Year Plan. The plan aimed to reduce disparities in income and wealth, to eliminate exploitation and to provide security to tenants, as well as to achieve social transformation through equality of status and an opportunity for different sections of the population to participate in development initiatives.

FARM NUMBERS, FARM TYPES, AND CONSERVATION PROGRAMMES

The number of farms has declined dramatically since its peak of 6.8 million in 1935, with most of the decline occurring during the 1940s, 1950s, and 1960s. The decline in farm numbers has levelled off since the 1970s. By 2002, 2.1 million farms remained. The remaining farms have a much larger average acreage, but averages mask differences among farms. Today's farms range from very small retirement and residential farms to industrialized operations with sales in the millions. Part of this diversity stems from the very low sales threshold ($1,000) necessary for an operation to qualify as a farm for statistical purposes.

One way to address the diversity of farms is to categorize them into more homogeneous groups. The farm typology developed by ERS identifies five groups of small family farms (sales less than $250,000): limited-resource, retirement, residential/lifestyle, farming-occupation/low-sales, and farming-occupation/high-sales. The typology also includes large family farms, very large family farms, and nonfamily farms. In addition, very small farms (sales less than $10,000) make up more than half of all farms. Very small farms account for a particularly large share of farms in the limited-resource (72 per cent), retirement (76 per cent), and residential/lifestyle (76 per cent) groups. Production, however, is concentrated among larger farms; small farms account for only 27 per cent of the total value of production.

The smallness of most farms has implications for conservation and the environment. An ERS study found that smaller corn farms are less likely to use conservation tillage than are larger farms. The practice is more practical for larger farms because they have more acres over which to spread the cost of new or retrofitted equipment necessary to adopt conservation tillage. Small farms whose operators are retired or farm part-time are also less likely to adopt conservation tillage, possibly because of hesitancy to change familiar production practices. Small farms, however, participate widely in the Conservation Reserve Programme (CRP) and the Wetlands Reserve Programme (WRP).

SIGNIFICANT ROLE IN FACILITATING THE PROCESS OF AGRICULTURAL DEVELOPMENT

A number of challenges face FSD if it is to continue playing a significant role in facilitating the process of agricultural development. Some of these are:

- Better Incorporation of Farmers, Incorporating farmers into the research process was one of the most important principles underlying the evolution of the farming systems approach. The basic justification for this was that farmers could improve the efficiency of the research process. At a minimum, it was argued they could prevent the use of research resources on types of technologies that they would be unlikely to adopt, while even more Importantly, it was anticipated they could, in their own right, contribute positively to the research process. Unfortunately, farming systems workers often have not sufficiently recognized this positive and interactive contribution of farmers. Consequently, this spawned the farmer participatory research (FPR) movement. The problem of not utilizing the farmer sufficiently in FSD activities is not due to inherent deficiencies in FSD itself but rather with the way in which it has been applied, Techniques of FPR or PRA need to be incorporated into FSD.

- Continued Evolution of FSD. FSD is relatively new. Therefore, the methodology is still evolving and, as a result, universally accepted 'standard texts' on the 'nuts and bolts, of how to do it are still to emerge. Related to the methodology issue is that time and cost-efficient (*i.e.*, money and people) methods for undertaking FSD still need further development. This is important because of the limited resources available for undertaking research, both on-station and on-farm.
- Greater Incorporation of the Policy/Support System Perspective. As has been stressed already' the farming systems approaches implemented to date have focused mainly on the technology dimension (*i.e.*, FSR). However, a basic principle of the FSD approach is that the farming systems perspective is critically important in formulating and adapting policy/support systems in ways that will facilitate and accelerate the agricultural development process, Once again, discussions relating to the policy/support system are scattered throughout the manual, and it is highlighted as an essential component of FSD. Failure to incorporate this dimension in FSD activities is likely to have an impact analogous to playing soccer on one leg, Nevertheless, because of the lack of much proven experience and documented material on the farming system perspective with respect to the policy/support system, this manual does not deal with this dimension to the extent that would be desirable. Hopefully, this lack of experience will be rectified in the fairly near future. An example of the potential value of using a policy/support perspective in FSD is given in Box 3.8.
- Incorporating Equity Issues — Intra and lnter-Generational. FSD tries to help the farmer with the problems he or she has identified. Of course, the reason for this is the necessity to introduce an intervention in which they are interested, It its likely that these 'felt, problems of farmers are likely to have a short-run focus (*i.e.*, particularly when the farmer is operating very close to the survival level). Also, it is possible that helping to solve the individual

farmers' problems creates others for the society as a whole. Although the example given in Box 3.9 pertains to inter-household relationships, equity considerations within a particular generation also apply to what is happening within farming families. As will be discussed later, it is wrong to assume that all farming households or families operate in such a way that distribution of effort and benefits is equitable. For a positive example of the benefits of understanding intra household relationships. When designing a technology to help farmers increase their productivity, consideration must be given to the possible long-term effects (*e.g.*, decreasing the amount of productive land available) of that intervention. Therefore, if farming system (FS) workers are not careful, their work can result in creating two types of inequalities, that is, helping some farming households — or even certain individuals within those households — at the expense of others and/or reducing the quantity and quality of land that can be productively farmed by future generations. Avoiding the development of such inequities constitutes a major challenge to FSD teams who need techniques to be able to screen out proposed technologies and policy/support systems that would encourage or exacerbate such trends. Once again, issues relating to equity matters, although generally recognized as important, still need attention in farming systems type activities. This will require continued evolution of FSD towards the FS 'in the large' approach mentioned earlier.

- Assessing Agricultural Research Impact. Related to the research resource issue is the importance of devoting some effort to assessing the impact of the research process-something that often has been done inadequately. More attention needs to be paid to adoption/ diffusion studies. Such studies, of course, are the best measure of the impact of the agricultural development process, Therefore, the credit for favourable results from such studies cannot be allocated to FSD activities alone. However, FSD teams, because of their on-farm location and pivotal linkages with other actors, are often in the best position to take a leadership role in their execution. Also, as suggested earlier, such studies can and should be used for other purposes, such as feeding back priorities for further research and providing evidence for adjustments in the policy/support system to encourage greater adoption,
- Improving Credibility of FSD. Establishing credibility for FSD-related activities is a major challenge and is necessary to ensure that some of the limited research resources always will be allocated to them, As argued earlier, FSD staff constitute only one set of the 'actors' in the agricultural development process. Additionally, FSD

helps facilitate a process and does not result in a product by itself. Thus, FSD cannot claim sole credit for any technologies developed for, or adopted by, farmers. However, it achieves credibility through its linkages and cooperative efforts with other 'actors' in the agricultural development process. These 'intermediate products' need to be documented and publicized to a greater extent than often has been the case.

IDENTIFY APPROPRIATE DEVELOPMENT STRATEGIES

On-farm systems studies were carried out in village areas south of Bangalore, India to determine the appropriateness of the current livestock policy. Farmers without irrigated crop land are the poorest in this area and at greatest risk from the effects of climatic and biological variations. These farmers depend on small livestock herds for cash income and on subsistence cropping for home consumption. When their small-scale livestock production fails, farmers are forced to seek outside employment for cash, and thus cropping activities suffer.

This farming system study identified two improvements that would contribute to better livestock productivity: control of epizootics and a 'preventative veterinary medicine service' focused on improved fertility management among small herd-owners. Previously, veterinary service policy had been to target larger herds where the immediate payoff had appeared to be much greater. Such a livestock policy change should lead to greater stability in the production of animal products for local markets; less reliance on seasonal employment and, therefore, greater stability of the household labour force; and more stable subsistence cropping activities.

An Intragenerational Issue

In Botswana, farmers have a tradition of sharing draught power, In this semi-arid environment, field operations such as ploughing and row planting depend on access to draught in a timely fashion (*i.e.*, days on which there is adequate soil moisture for ploughing/planting). If a technology system is introduced that requires more draught, rather than less, it may mean that the household controlling the draught may not be able to routinely allow other households to also use their animals under a sharing arrangement. Thus, helping one household may harm other households in the community.

Decision Making and Trade-offs

An FSD programme in Amphoe Pharao, Thailand discovered a positive and unexpected benefit from interaction at the intra-household level. The introduction of improved and more commercially oriented poultry operations was targeted towards women members of households, because they traditionally manage and derive income from poultry. However, improved technological interventions not only required new inputs by men but were discovered to provide profitable use of under-employed male labour in many households. This example demonstrates decision making with interactions between members of the household that are desirable for all: a win-win situation. Often, however, decisions for change favour one segment of the household, while adversely impacting on another.

COMPARING STATION-BASED RESEARCH AND FSD

The literature frequently compares and contrasts station-based research and FSD. The tone in such comparisons often implies that they are substitutes for each other. However, both station based research and farm-level FSD are needed. This is because they focus on different things that are complementary to each other.

A major difference between them is the following:

- On the experiment station, applied research is usually undertaken, in which new technologies are created.
- FSD, on the other hand, concentrates mainly on adaptive research, which involves helping to adjust technologies to specific environmental conditions. FSD also helps feed back information about future priorities for applied research to experiment stations.

Some of the major differences between station-based research and FSD and helps indicate why both are necessary, particularly in areas where the Green Revolution has not occurred. In such areas, greater attention will need to be paid to adaptive research, if relevant improved technologies are to be developed and adopted by farmers. Obviously, strong linkages between these various contributors are of crucial importance. The FSD approach, involving working directly with the farmers, is a more 'bottom-up' or 'micro to macro, orientation, compared with the more 'topdown, or 'macro to micro' orientation of research work or planning exercises that start at the experiment station or the upper levels of planning ministries. Part 'B' indicates that FSD can help in strengthening 'bottom up' linkages amongst the various groups. In doing so, it helps in the process of improving productivity but does not result in a product by itself. (*i.e.*, facilitates a process rather than producing a product).

The relevant improved practices/technologies and policy/support systems are needed to bring about increased agricultural productivity. Decisions on which strategy is to be emphasized to increase agricultural productivity will depend on circumstances. For example, if no technology is available to ensure an economic return to fertilizer, there will be little value in concentrating on developing an input distribution system that makes fertilizer available to farmers. Therefore' the complementary nature of the relationship between developing and disseminating relevant improved technologies and policy/support systems needs to be constantly borne in mind, if the agricultural development process is to proceed efficiently.

It is apparent that the FSD approach has a broader focus than simply addressing technological issues — the primary focus of early FSR activities. The FSD approach, can help also in facilitating linkages not only between farmers and station-based researchers but also with other 'actors,, including those responsible for designing and implementing the policy/support system. In fact, most work in the farming systems arena to date has concentrated on the technology thrust or farming systems activities relating to FSD. This manual concentrates most attention on the research or technology thrust of FSD although linkages with other 'actors' besides farmers and station-based researchers are discussed to some extent.

CONCEPTS IMPORTANT IN THE FARMING SYSTEMS APPROACH TO DEVELOPMENT

To design appropriate or relevant ways of helping farmers, it is essential to understand the conditions under which farmers are operating. They have in fact, very complicated farming systems. The factors that have an influence on what the farming system will be. The operator of the farming system is the farmer or the farming family. To farmers, the way in which they earn their living and the economic, social, and cultural well-being of their households are linked closely and cannot be separated.

The members of the farming household have three basic types of inputs: land, labour, and capital. Management involves allocating these to three different activities or processes, that is, crops, livestock, and off-farm enterprises. In making decisions on how to allocate their inputs in producing one or more products, farmers have to make some difficult decisions. These decisions will involve using their knowledge to come as close as possible to fulfilling the goal(s) for which they are striving. These goal(s) may vary from farmer to farmer (*e.g.*, maximizing their income, producing enough food to feed the family, etc.). Many farmers are likely to want to maximize their incomes, once they have made sure they are producing enough food to feed their family and have met other societal obligations. The resulting combination of products (*i.e.*, crops, livestock, and off-farm enterprises) they are producing with their inputs results from the farming system they have adopted. Nevertheless, the extent to which the farming system fulfils the goal(s) they have chosen will depend on the managerial skills of the farming family and its ability to make good decisions in the allocation of inputs in a very uncertain production environment.

However, some parts of the environment that influence what the farming system will be are outside the control of the individual farming family thus causing uncertainty as far as the farmer is concerned. The 'total' environment in which farming households operate consists of two parts: the technical (*i.e.*, natural or physical) element and the human element.

- The technical element determines the types of, and physical potential of; livestock and crop enterprises. For example, in some areas, the low level of rainfall allows sorghum, but not maize, to be grown on rainfed land. The technical element includes physical and biological factors. These often are modified to some extent through technology developments — for example, increasing water availability through irrigation Improving soil quality by adding fertilizer, breeding for yield stability during drought, etc.

- The farming system that actually evolves, however, is only part of what is potentially possible. The human element is important in determining what the actual farming system will be. The human element consist of two types of factors: exogenous and endogenous.

Exogenous factors (*i.e.*, the social environment) are largely out of the control of the individual farming family. These factors will influence what the farming family can do and can be divided into three broad groups:

- Community structures, norms, and beliefs.
- External institutions, which include extension, credit, and input distribution systems on the input side and markets on the output side.
- Other influences, such as population density, location, and infrastructure.

Endogenous factors, on the other hand, are those the individual farming household controls to some degree. These include the types of inputs mentioned earlier, that is, land, Iabour, and capita]. It is important to recognize that these resources and managerial ability vary among households and regions. The resources vary on the basis of quantity and quality, both of which influence the performance and the potential of the system. In addition, these inputs or resources may or may not be owned by the household. Access to one or more of these resources may be on another basis of use (*e.g.*, borrowing draught animals), which may limit or restrict the ease or intensity of use and thus, in turn, affect the goals and performance of the farm family.

Nevertheless, it is the fanning family that decides on the farming system that will emerge. However, this system will be influenced and sometimes constrained by the technical element and exogenous factors.

The farming system is obviously complex, and the results can vary greatly because of differences in the 'total, environment. These facts help explain why some technology thought to be relevant often has not been adopted, or when it has, why the degree of adoption has varied widely. Not considering the human element in agricultural research has contributed to many socalled 'improved' technologies being irrelevant.

Status, Goals, and Appropriate Technology

Typical limited resource farm households in low income countries generally are characterized as follows:

"Farm households, with access to their means of livelihood in land, utilizing mainly labour in farm production, always located in a larger economic system, but fundamentally characterized by partial engagement in markets which tend to function with a high degree of imperfection."

Important points to note about this definition are that:

- Such households often are subordinate to some other external forces. That is, they do not have complete control over their own destiny and are in a process of transition. This is because of not being integrated fully in the market economy.
- They have access to land to pursue their livelihood. They cultivate the land largely with family labour, in conjunction with only small amounts of capital.
- The input or factor markets (*i.e.*, land, labour, and capital) work poorly. Concerning land, non-market rights of access or non-price forms of tenure are more likely to operate than a freehold market. The capital markets usually are developed poorly and variable production inputs are often not available. As a result:

"Credit and interest rates may be tied to other factor prices like land and labour within a dependent economic relationship. Thus factor markets may be locked together contractually rather than being independent."

MARKET INFORMATION

Also, market information may be highly imperfect (*i.e.*, not available or readily accessible). Thus, it is not surprising that sharing and reciprocity often exist between households. In these non-market transactions, exchange of unlike goods and services take place, which are not valued by market prices. Sometimes, such relationships exist between households that are related to each other.

- Another reason why such households are integrated only partially into the market economy is because they consume a proportion of the product they produce. This enables them to have some ability to survive independently of the larger system, which may be important in explaining their economic behaviour.

  A number of microeconomic theories have been developed in an effort to explain the economic behaviour of farm households. These are based on various assumptions about household goals and the characteristics of the markets within which households make their decisions. Ellis [1988] notes that these theories are not mutually exclusive, sharing as they do certain key assumptions such as:
- The household is a single decision-making unit for economic analysis purposes. As a result, it maximizes a single utility (satisfaction) function that represents the joint welfare of its members.

- Profit maximization and utility maximization coincide where income is the only variable in the utility function. Profit maximization is always one of the components of utility maximization" when all the input and output markets are fully formed and competitive.

Ellis [1988] suggests that the theories predict different results because of different assumptions about the working of the factor and product markets, rather than because of differences in assumptions about household goal(s). In fact, an important characteristic distinguishing many of the theories results from different assumptions about labour markets and the allocation of household labour time, In support of these assertions, Ellis then discusses a number of these theories of economic behaviour. He calls these profit maximization, risk aversion, drudgery aversion, farm household theories, and share tenancy. In doing so, he recognizes the impossibility of separating short-run household decisions from the wider social relations of production.

Consequently, there is likely to be a certain universality in the aspirations of limited-resource farming households. The most important elements are income, effort avoidance, and risk avoidance. In economic terms, this means such households try to increase their utility or satisfaction, which increases with income but decreases with greater effort or higher levels of risk. This can be restated as maximizing income for any given level of effort or risk. Alternatively, it can mean reducing risk, or effort, for a given level of income.

Attempts to maximize utility or satisfaction take place within a set of constraints. The constraints, rather than the aspirations, lead to the most important differences in farming systems. Some of the determinants of the farming system. It is easy to visualize constraints relating to the technical element causing differences in the type and productivity of farming systems, However, also important in the differentiation process are the factors relating to the human element. The quantity, quality, and ratios (*i.e.*, particularly between land and labour) of the endogenous factors (*i.e.*, factors of production — land, labour, and capital) play an important role in further differentiating farming systems. The extent to which they can be modified will depend on the degree of integration into the factor markets (*i.e.*? input markets). Also the products resulting will be influenced to some extent by community norms and beliefs and the degree of integration into the product market.

The important implications for what is appropriate technological change. For example, limited resource households often are likely to be very little above survival level and, hence, will place a very high premium on improved stability of production. The levels, qualities, and relative proportions of the factors of production are important in indicating the most appropriate route for improving the productivity and profit of the farming system — which is maximizing the return to the most limiting factor. For example, in areas of high land/labour ratios (*i.e.*, low population densities), labour saving (*e.g.*, mechanical, extensive livestock production) types of technology/systems are likely to be most appropriate. In contrast, in areas of low land/resident ratios (*i.e.*, high population density) land saving technologies/systems (*e.g.*, biological and some chemical, intensive livestock production) will likely be more appropriate. However, increasing the return to the most limiting factor can have an indirect positive or negative impact on the use and productivity of other inputs. Thus, in view of the imperfect operation of the factor markets, care has to be taken in evaluating what would be the most relevant technologies.

For example, in a semi-arid area, initial superficial examination may indicate that because of the high population densities, strategies designed to increase the productivity of land would be most important. However, given the seasonal nature of agriculture in semi-arid areas, there will be periods of intense labour activity during the year. These will create labour bottleneck periods, followed by periods of underemployment. Thus, land-augmenting strategies using labour at critical labour bottleneck periods — when the opportunity cost of labour is high (*i.e.*, there are good alternative uses for labour that can yield an equally high return) — could be unattractive to the household. This is likely if adoption of such strategies involves the use of more labour at such labour bottleneck periods. It will also be likely if the productivity per unit of labour applied at that time is higher in alternative uses.

The results indicated the superiority of the improved maize technology over both improved or indigenous practices for sorghum and cotton in terms of both return per unit of land and labour. Note that for both the sorghum and cotton improved technologies, the return per unit area was higher than that from employing indigenous practices; but that the return per unit of labour used during the June/July period was the same or lower. Unlike the sorghum and cotton improved technologies, the improved maize technology was, from a profit viewpoint, suitable for farmers whether they were faced with a labour or land limitation.

Not surprisingly, since the 1970s, improved maize technologies have been adopted widely in Northern Nigeria, in a mini-Green Revolution. Although, initially the prime motivation for growing maize was as a cash crop, it eventually became a more important dietary item as farming households made trade-off decisions between growing the high yielding maize rather than the low yielding sorghum.

FARM PLANNING

Planning means taking decisions in advance. It stimulates thinking, broadens understanding & challenges the farmer to move forward. It is a forward-looking approach. The farm plan helps a farmer to decide how to combine new ideas & old ones to his best advantage. By identifying his credit & supply needs, the farm plan helps him to arrange for the timely supplies of credit, seeds, fertilisers, etc. A specific farm plan setting fort his expected output, expenses & income, serves as a sound basis on which a credit institution can give him production credit, based on his productive capability rather than on his net financial assets. It is out of his income & not through the sale of assets that the cultivator has to pay off his loan. Thus the farm plan or the budget is to the farmer what the blue-print of the architect is to a building contractor. It shows what is to be done & how it is to be done. It furnishes an organised & logical approach to his problems & helps him to work out the solution.

MEANING OF FARM MANAGEMENT

Farm Management is the application of business principles and the scientific principles of agriculture (as discovered by the chemist, the physicist, the agronomist, the animal husbandmand and other specialists) to the business of farming. The farm manager is the one who plans the farm, arranges the cropping scheme, provides the equipment and actually transacts the business and sees to the operation of the farm.

The term is a new one and it is important that students of farming get a clear understanding of its meaning. Because of the close relation of farm management to soils, farm crops, live stock, and other agricultural subjects, people often confuse the terms and gain erroneous ideas of what farm management really is.

Farm management is also related closely to rural economy. There is danger of confusion arising in the consideration of these terms also. Questions which relate to the single farm as a unit or to a group of farms operated under the same organization are matters of farm mangagement. Those which relate to a group of farms operated separately, or to a community, are matters of rural economy. The farmer must meet both classes of questions. To state it differently, those things that have to do with the organization and operation within the individual farm relate to farm management. Those that have to do with the external relations of the individual farm to the group or community, relate to rural economy.

The Relation of Farm Crops to Farm Management

The farmer must grow crops on his land either to sell or to feed to his live stock. This makes it necessary for him to know how to prepare the land, sow the seed, cultivate the crops and harvest and store the product. In gaining this knowledge he studies the questions from the viewpoint of the agronomist and considers only what will be necessary to provide the best conditions for the growing crop.

The farmer must also decide what crops he will grow. One farmer will want to grow corn, another will grow cotton, and still another will grow hay, wheat or potatoes. Usually he will grow several different crops. He will need to decide how much of each one should be grown. In deciding these questions he will study the questions from the viewpoint of the farm manager. As an agronomist, the farmer is interested only in how to grow the crop. When the crop is mature and harvested, the problem is completed. As a farm manager he still has to settle the question of what to do with the crop. It may be sold. Would it pay better to feed it to live stock? If so, what kind of live stock will give the best returns for the feed used? These are farm management questions.

Relation of Animal Husbandry to Farm Management

Most farmers grow some live stock. They must know the nature of the animals they wish to grow, the kind of food they eat, how to prepare it for them and when to feed them. They should also know how to select the best animals and to what purpose the various classes of animals are best adapted. In answering these questions they study them from the viewpoint of the animal husbandman and cosider only what is best for the animals.

The farmer must decide, however, what classes of stock he will grow. On some farms dairy cows will be best, on others beef cattle and hogs will be a desireable combination.

There are farms on which horses and sheep with some poultry will be preferable. There are also the question of how much live stock the farm will supply feed for, or what to do with the surplus when more is produced than is required for the live stock. In answering these questions they are studied from the view point of farm management.

Relation of Rural Economy to Farm Management

The management of a farm includes the buying and selling of products. It is important that the farmer have a knowledge of markets and of values, and that he know when to buy and sell. Often cooperative marketing offers the best opportunity for disposing of products at a profit. So long as the farmer is concerned only with the business enterprises relating to his own farm, he is within the field of farm management. When the business activities and social welfare of several farms or an entire community are collectively considered, the question becomes one of rural economy. Farm management deals with the problems of the individual farm from the business point of view. Rural economy deals with the problems of a collection of farms or an entire community.

Studying Farm Organization

The student of farming is taught how to grow crops by the agronomist, how to grow live stock by the animal husbandman. He learns how to know insects and combat them from the entomologist, and how to cooperate with his neighbors in selling his crops or his stock, from the rural economist. Finally, from the viewpoint of a farmer, he will need to select a specific farm, make plans for its arrangement into fields, select the kind of stock and crops that it will be best to raise, and determine how much of each should be raised. He must provide the buildings and equipment, estimate the capital required and the labour necessary to operate the farm. The farm is thus organized into business form and its operation is simplified. This art of farm organization is learned from the farm manager.

Meaning Made Clear by Questions. "How shall I grow corn?" is clearly and agronomy question. How to select and grow cattle, is just as clearly a question for the animal husbandman. "Shall I grow corn? or some other crop? or cattle? Having grown corn, shall I sell it? or feed it to cattle? or hogs?" are farm management questions. If questions similar to the above are put to each problem presented there should be no difficulty in distinguishing the viewpoint from which it should be attacked.

Definition of Terms. In the pages following, reference will be made to the terms, "Capital," "Inventory," "Farm Receipts," "Farm Expenditures," "Farm Income," and "Labour Income." An explanation of these terms is given below.

Capital. Economists divide capital into two classes: (1) Fixed capital, including land, buildings and permanent equipment, as teams, implements and live stock for purposes of production; and (2) circulating capital, which includes values in seed, feed, fertilizers, supplies and market stock or crops as well as cash for operating the farm. The term is limited to those forms which disappear in the process of production, reappearing in other materials of similar kind.

There is a tendency on the part of farmers to tie up most of their capital in fixed forms, leaving too little for circulating or operating capital. The Inventory is a detailed list of the farm property with values assigned. An average of the values at the beginning of the year and at the close is considered to be the value of the capital invested in the business. The difference between the opening and closing inventories shows the increase or decrease in the value of the farm property.

Farm Receipts are derived from two sources: (1) direct and (2) indirect. The direct receipts are represented in the cash received from the sales of crops, live stock, live stock products, for labour for others and from miscellaneous sources. The indirect receipts would arise in an increase in the inventory due to permanent improvements or to appreciated values of any of the various forms of property. Values should not be inflated for the purpose of showing a gain in the inventory. Debts cannot be paid on anticipated rises in land values or in values of property unsold.

Farm Expenditures

Items paid for in cash for the support of the farm only can properly be classed as farm expenditures. Items of personal or family expense should not be charged against the farm in determining the profits from farming. A decrease in the inventory is considered as an indirect expenditure. The amount of the decrease should be deducted, with the direct expenditures, from the receipts to show the net farm income. The Farm Income is the difference between receipts and expenses. It is the return for the use of capital and unpaid labour. The farmer and his investment unite in earning the farm income. To learn what is earned by the unpaid labour only, the amount that the value of the investment would earn if placed at interest, must be subtracted from the farm income.

The Labour Income is the amount earned by the farmer alone. To learn what this amount is, the amount earned by the money invested at the local interest rate and the value of the labour of other members of the family at local wages while employed at labour on the farm, must be deducted from the farm income. The balance will be the labour income of the farmer.

INTEGRATED FARMING SYSTEMS

Development and adoption of integrated farming system provides high opportunities of productivity enhancement, employment, income generation and nutritional security by diversifying and integrating different components of farming, *viz.* crops, horticulture, livestock and fisheries (depending upon location specificity). The systems based on multiple recycling of carbon, energy and nutrients would also help minimize environmental loading with pollutants. The different components of the system have complementarities with waste products of one component becoming source of food and energy for another.

The researchable issues encompassing analysis of nutrient and energy fluxes among system components, path analysis of bio-physical constraints, multiple uses of water, water harvesting and recycling, socio-economic conditions of people, internalization of ITK and market analysis, etc. are to be addressed through basic and strategic research by ICAR institutes and SAUs. The integrated technology would be developed integrating crops, horticulture, fisheries & livestock by ICAR Research Complexes having R&D facilities on different system components at SAUs. Further assessment and refinement of developed farming systems to cater to region and location specific requirements would be accomplished through centres of AICRPs on Cropping/Farming System Research and SAUs.

The technology would be disseminated through KVKs, SAUs, NGOs, SHGs, etc. with true participation of ATMAs and different functionaries of State Line Departments (Agriculture, Horticulture, Animal Husbandry, Fishery) at district and block levels. The backward and forward linkages between KVKs and state functionaries would provide the required feedback to SAUs and ICAR institutes on upgradation of technology. The KVKs would organize regular trainings on different aspects of technology to extension personnel and various stakeholders.

ESTABLISHMENT OF AGRI-INDIA KNOWLEDGE PORTAL: A SINGLE ELECTRONIC GATEWAY

The information on improving agricultural productivity and protection of crops, livestock and natural resources from damages caused by disasters and unsustainable activities is focal to the rural farming communities. This would require access to a wide range of information on new technologies, alternate varieties of crops, improved breeds of livestock, information related to soil, water quality, information on pesticides, farm implements, animal health, weather and other related aspects. Besides farm-activities, non-farm activities also constitute sources of income and livelihood for the rural farmers. Therefore information is also required on the means to produce food products required by markets and ability to sell them as also the markets, cold-chains, warehouses, processing and other avenue.

ICAR-ERNET network consists of 274 Institutions including National Agriculture Institutes, State Agricultural Universities including some of their colleges and Research stations, National Research Centres, Project Directorates connected through Leased line and VSATs. Besides, a VSAT-based network of Krishi Vigyan Kendras could be had. Both these networks will be linked through a high capacity leased line to make a unified high speed secured Intranet. Since the intranet of NARS institutions is built over the existing ERNET network, it is planned to host the core components of knowledge portal application at the ERNET Hub.

The Department of Telecommunications has set up several sub-groups on network expansion, broadband, telecom equipment manufacturing and R&D to examine various options to enhance growth. The sub group on network expansion has targeted that subscribers numbers must cross 250 million by 2007 and 500 million by 2010. Another target is mobile coverage of 85 per cent of the country by the end of 2007. To boost the broadband penetration in the country, another sub-group will examine the ways to provide broadband coverage for all panchayats by 2010 and all secondary/ higher secondary schools and public health care centres by 2007. The Portal application will be constructed and deployed using suitable Portal management software.

The core functional components of the portal application software are:

- User Registration: The user registration process is a mechanism for acquiring the information related with the user of portal which is an integral part of the portal.
- User Authentication: The user ID and the password pair generated as end result of registration process are authenticated through a carefully designed mechanism and form a basis for accessing the various resources of a portal.
- User Access Control: Once the users are authenticated, an access control mechanism built into the portal determines which data and

information can they see/manipulate, and how will it be displayed? The User Access Control mechanism is designed based upon the specific role that the particular user will perform in the portal.

- Personalization: Personalization is a feature that enables a user to access targeted web content that meets a user's needs and preferences from the enormous web content on the portal.
- Content and Document Management: Content and document management system in the portal application streamline redundant content development, publishing and management processes to increase operational efficiencies, and improve content quality.
- Document/workflow: The various rules to manage the workflow and lifecycle of documents are established in a portal, which automates the manual, labor-intensive process and results in consistency and auditability in authoring and producing content.
- Collaboration: The portal applications facilitate collaboration among the various stakeholders/ partners via online chat rooms, discussion groups, online document sharing, messenger applications, etc. The digital content of the portal will be developed through a peer review process. For this purpose, 15 Content Accreditation Centres, *i.e.* one each in the 15 Agro-climatic regions of the country could be had. Each accreditation centre will coordinate with other SAUs and agricultural institutions in their region for development of content in regional language as well as in English and also do its validation, which will be collected in the central data warehouse integrated in the knowledge portal.

The intranet network consisting of various stakeholders primarily ICAR institutes, SAUs and KVKs will facilitate the development of Central Agriculture Data warehouse, Content Development and Validation System, Agricultural Best Practices, Expert System for users, Multi-lingual content development and convertibility. The portal will also serve as a platform for facilitation of interaction among researchers and extension workers in KVKs through high speed secure intranet.

Various applications and services of the knowledge portal will be accessible to the users including farmers, entrepreneurs, private sector organizations, other participating government and non-government organizations, extension workers, etc. through Internet. The services include Agro-met advisory services, Market intelligence, Packages of practices, Agri-business consultancy, e-learning, knowledge on indigenous farming practices, and agriculture related FAQs. Besides, it will also act as information gateway for all agriculture related schemes and programmes etc. The Portal will also serve as a gateway for up-to-date websites of various agricultural research and education organizations accessed through Internet. The portal will provide a platform for facilitation of interaction among the end user communities through Internet also. The Agri-India Knowledge portal will also provide for collaboration among other agencies related with agriculture like Meteorological Dept, Agricultural Markets, Banks and other financial organization, Input suppliers, Agricultural equipment manufacturers, Non-agricultural universities for providing services through Agri-India knowledge portal both through secure and high speed Intranet and Internet.

The services proposed to be implemented in the collaborated mode include On-line education, Information on Market Prices, Weather Information including early warning system, Agricultural Input status information, Credit and crop insurance services, land record information, etc. The AIK Portal is neither intended to replace any agency websites nor any other government portal. This Portal will, however, provide additional means by which those sites might be accessed by a possibly wider and more varied client base. The AIK Portal will open opportunities for multiple agencies to participate in web delivery and development opportunities.

The Portal would be able to integrate diverse interaction channels at a central point, providing a comprehensive context and an aggregated views across all information related to agriculture including *know-what* or declarative knowledge, *know-how* or procedural knowledge, and *know-why* or usual knowledge and *creation of new knowledge*.

FARMING SYSTEMS APPROACH TO DEVELOPMENT

Details concerning the characteristics and methodology of the FSD approach, as developed by the Farm Management and Production Economics Service of FAO, are available elsewhere. Thus, the following discussion represents only a brief summary. The primary objective of FSD is to improve the well-being of individual farming families by increasing the overall productivity of the farming system in the context of both the private and societal goals, given the constraints and potentials imposed by the factors that determine the existing farming system. It is based on the development principles of improving productivity, increasing profitability, ensuring sustainability, and guaranteeing an equitable distribution of the results of production,

The farm household is the principal system and focus of FSD and consists of three basic subsystems, which are closely interlinked and interactive: the household, the farm, and off-farm activities. The two major categories of activities or thrusts of FSD, both of which involve intensive interaction with farmers? are:

- Farming Systems Analysis. This involves studying, together with the farmers the natural (*i.e.*, technical) and socio-economic (*i.e.*, human) environments in which farm households operate. The aim is to identify the constraints limiting farm productivity and production and hindering improvement in the welfare of the farm households themselves. Potential solutions to these problems are identified, and the results of this analysis, formulated as suggestions for further action then are passed on to the relevant 'actors'. These could include researchers, extension and support service staff, and/ or policy makers.

- Farming Systems Planning, Monitoring, and Evaluation. These involve testing, monitoring, and evaluating improvements on-farm, with the direct involvement of farmers. Examples of such activities include those regarding proposed technological improvements, proposed revisions in farm plans, and improvements in support services and farm-level impact of proposed policy changes. Those improvements thought to be potentially useful then are disseminated to other farmers via the extension service. Prior to this stage, it may be necessary to negotiate adjustments in the policy/ support services that will facilitate the dissemination and adoption of the improvement. After dissemination, it is of primary importance to monitor and evaluate the adoption rate of the proposed improvements by the farming community. This provides an indication of the actual value of the improvement being disseminated.
- Many individuals in the last 15 to 20 years have written extensively to clarity the concepts of the farming systems, particularly with respect to the development of relevant improved technologies. There are many ways of presenting a diagram involving the application of farming systems methodology to research (*i.e.*, technology development). One such way of emphasizing the technical research aspects of FSD. There are four fundamental stages in the FSD approach.
- The Descriptive or Diagnostic Stage, in which the actual farming system is examined in the context of the 'total, environment — to identify constraints farmers face and to determine the potential flexibility in the farming system in terms of timing, unused resources, etc. An effort also is made to understand the goals and motivation of farmers that may affect their efforts to improve the farming system,

  The following sequence of activities is usually undertaken:
  * After deciding on the location of the work, start-up activities include reviewing secondary sources of information, making the necessary contacts, assembling the professional team, and making logistical arrangements,
  * Farming families are classified tentatively into homogeneous groups. The farming families within each group or domain usually practice the same farming system(s), face the same

constraints, and have the same potential solution(s) to their problems. This tentative classification can be modified as additional information becomes available.

* An informal exploratory diagnosis or survey is undertaken to obtain a qualitative understanding of the determinants of the existing farming systems (*i.e.*, both biophysical and socioeconomic). At the same time, efforts are made with farmers to ascertain the constraints, flexibility, and potential opportunities in the farming systems they are using currently.
* Sometimes a verification survey involving a structured formal survey, suitable for statistical analysis, is undertaken to quantitatively confirm insights obtained in the exploratory diagnosis. It also can be used to provide a database for farm and development planning, policy analysis, monitoring, and evaluation.

- The Design Stage, in which a range of strategies is identified that are thought to be relevant in dealing with the constraints determined in the descriptive or diagnostic stage.

The usual sequence of activities involves the following:

* Together with the farmers, the constraints are ranked according to their severity, and potential solutions are identified after determining what flexibility exists in the farming systems currently practiced. Information for designing such strategies (*i.e.*, particularly relating to technological issues) comes from experiment station work, researcher managed and researcher implemented (RMRI) type work on farmers' fields, and from other farmers.
* An evaluation is made of the proposed solutions before putting them into practice. These solutions can be technological or institutional in nature. A number of analytical techniques can be used to evaluate the potential technical feasibility? economic viability, social acceptability, and ecological sustainability of the proposed solutions before they are put into practice. One or more of the proposed solutions are selected for actual evaluation on-farm.

- The Testing and Implementation Stage, in which one or more promising strategies, arising from the design stage, are examined and evaluated under term conditions to determine their suitability for producing desirable and acceptable changes in the existing farming system.

Activities at this stage involve the following:

* On-farm testing or evaluation with farmers determining how

well the potential improvements fit into the system, whether or not they are acceptable to farming households, and what modification may be needed to make them acceptable. In terms of technological issues, this stage conventionally has consisted of two steps:

— Researcher managed but farmer implemented (RMFI) tests to establish whether previously determined technical relationships are altered by farmers' management of non-treatment variables.

— Farmer managed and implemented (FMFI) tests, when the team is confident that technical relationships will hold but needs to evaluate the proposed technologies under local socio-economic circumstances. In addition to the on-farm evaluation of the proposed technologies, evaluations also can be made of the proposed farm plans and of proposed changes in support systems and/or policies.

* Positive results of such evaluations provide justification for the FSD team to advocate further action. The tested technologies can be disseminated through the extension service to other similar farming households, The same applies to the farm plans that have been tested successfully. Favourable test results for proposed changes in support services and/or policies provide valuable farm-level information on necessary programme or policy adjustments (*e.g.*, in extension, marketing inputs and products, pricing policy, credit, etc).

- The Dissemination and Impact Evaluation Stage, in which the strategies that were identified and screened during the design and testing stages are extended to farmers. In terms of activities at this stage, impact/adoption studies can be very important. These potentially can be very useful, not only in giving some idea of the impact of, for example, agricultural research, but also in giving some idea of future priorities for agricultural researchers and indicating what adjustments are required in the policy/support systems to ensure better rates of adoption. Thus, through monitoring and evaluating the impact and rate of adoption of the changes that have been implemented and proposed/tested earlier, it is potentially possible to provide some indication of further desirable activities by both researchers and planners.

There are often no clear boundaries between the various stages, Design activity, for example, may begin before the descriptive and diagnostic stages and may continue into the testing stage, as promising alternatives emerge from RMFI trials. Similarly, testing by farmers may mark the beginning of dissemination activities. Also, going through all stages may not always be necessary. FSD team confidence in transferability during design/planning activities can sometimes mean going straight to FMFI work or even to recommendation/dissemination activities.

Thus, the process of FSD is recognized as being dynamic and iterative, with linkages in both directions between farmers, researchers, extension staff and policy/support service staff, The iterative characteristic can improve the efficiency of the development process by providing a means of identifying and fine tuning improved technologies for a specific location-that is, climatic situation, soil type, and/or farmer resource base.

Index